Retopia

Retopia tells the story of social innovation in times of crisis, and through its cross-disciplinary narrative it goes beyond existing forms of future anticipation and maps out a practice-based approach to the creation of new realities. It explores how new imaginaries, social experiments, and laboratories of societies can create spaces of possibilities, revalidate the peripheries, and create new forms of social coherence.

The peripheral regions in Europe are facing a crisis triangle: depopulation, the rise of the 'useless' class, and outdated social welfare systems. It is a crisis of political imaginaries and a lack of inspiring political stories. In response to this, the book specifically focuses on the concept of 'retopia', the idea of creating inclusive spaces of social innovation that encourage active participation. Through the creation of relocalized societies with a high degree of autonomy in 'leftover' spaces, such as Sicily, Western Latvia, or Northern Bulgaria, retopian redevelopment schemes offer new perspectives on 'ruined spaces'. *Retopia* uncovers the common links and limitations of utopian studies, future studies, degrowth, narratology, the commons, and political geography.

Retopia: Creating New Spaces of Possibility is an articulation of the potentialities of social innovation, political imaginaries, and future images, provoking a stimulating discussion among scholars and students in the fields of Politics and Future and Anticipation Studies.

Dirk Hoyer is Associate Professor at Tallinn University, Estonia. He has published numerous articles and a book on political imaginaries and retopia. Hoyer is also a writer, filmmaker, and script consultant.

Routledge Research in Anticipation and Futures

Series editors: Keri Facer and Johan Siebers

The Promise of Nostalgia
Reminiscence, Longing and Hope in Contemporary American Culture
Nicola Sayers

Indigenous Futures and Learnings Taking Place
Edited by Ligia (Licho) López López and Gioconda Coello

Sustainable and Democratic Education
Opening Spaces for Complexity, Subjectivity and the Future
Sarah Chave

A Creative Philosophy of Anticipation
Futures in the Gaps of the Present
Edited by Jamie Brassett and John O'Reilly

Anticipation, Sustainability, Futures and Human Extinction
Ensuring Humanity's Journey into the Distant Future
Bruce E. Tonn

Ethical Humans
Life, Love, Labour, Learning and Loss
Victor Jeleniewski Seidler

Working with Time in Qualitative Research
Case Studies, Theory and Practice
Edited by Keri Facer, Johan Siebers and Bradon Smith

Retopia
Creating New Spaces of Possibility
Dirk Hoyer

For more information about this series, please visit: www.routledge.com/Routledge-Research-in-Anticipation-and-Futures/book-series/RRAF

Retopia

Creating New Spaces of Possibility

Dirk Hoyer

Routledge
Taylor & Francis Group
LONDON AND NEW YORK
earthscan
from Routledge

First published 2023
by Routledge
4 Park Square, Milton Park, Abingdon, Oxon OX14 4RN

and by Routledge
605 Third Avenue, New York, NY 10158

Routledge is an imprint of the Taylor & Francis Group, an informa business

British Library Cataloguing-in-Publication Data
A catalogue record for this book is available from the British Library

ISBN: 978-0-367-25514-5 (hbk)
ISBN: 978-1-032-48387-0 (pbk)
ISBN: 978-0-429-28818-0 (ebk)

DOI: 10.4324/9780429288180

Typeset in Sabon
by Deanta Global Publishing Services, Chennai, India

Dedicated to:
kes

Contents

Preface

Starting to write a book about spaces of possibility in the middle of a global pandemic-related lockdown can be seen as either escapist and mentally confused or hopelessly idealistic. In addition, there is the scope of the book: find a new idea to reorganize (at least a segment of) society. Before even typing the first word, it was clear that this would be an uphill battle.

The prevailing credo: society-wise, it is impossible to find spaces of possibility. The world is stuck, not only in times of Covid lockdowns but also in terms of political imagination and new political narratives. With billions of people glued to their screens, the horizon has shrunken to nine inches. Finding spaces of possibility, or a retopia, is a mental exercise for pretentious people on the edge of swellheadedness. Except: it isn't. Moving the imaginary reconstruction of society into the field of either megalomania, conceitedness, or into the realm of the aloof is, in itself, a powerful control mechanism. It is time to undo this mental lockdown.

Of course, writing this book in an actual lockdown, with children in home schooling, increased workloads, and the miniaturization of other people to tiny tiles on a zoom screen has increased the challenge to escape the mental lockdown. Even before the omnipresent crises – global warming, Covid, the war in Ukraine – the future was already downsized to four words: There is no alternative. Innovation is celebrated in the business world, but social innovation happens in the margins. The many remarkable people who work on improving the setup of society arrangements on a small scale are operating in quasi-obscurity. While the world media writes eulogies about a new function of the 14th version of a mobile phone, very few people know about the Seikatsu Club.

Initially, the scope of the book was conceived as global, but the travel restrictions and the Covid-related limitations to everyday life have shrunken the spatial scope. Of course, there is a sad irony to this, especially in a book with "spaces of possibility" in the title. The book now has a largely European scope. At best, colleagues from the Global South take this as encouragement to provide a larger perspective.

This book is an incomplete quest, and if the outcome should be flawed, I hope it can at least encourage other people to embark on the journey out of the mental lockdown.

1 Political Agency in the Age of Downsized Expectations

In 843, the Kingdom of the West Franks was established, and most historians agree that this gave birth to the title of 'King of France'. With some interruptions France had kings (and emperors) for 1027 years, altogether 50 monarchs were sitting on the French throne. Over 1000 years of monarchy means over 1000 years of (imaginary) legitimacy. David Graeber reflected on the power of conviction and the establishment of belief:

> If you managed to convince everyone on earth that you can breathe under water, it won't make a difference: if you try it you will still drown. On the other hand, if you could convince everyone in the entire world that you were the King of France, then you would actually be the King of France. (In fact, it would probably work just to convince a substantial portion of the French civil service and military.)[1]

So, while breathing underwater (without the necessary gear) remains an illusion, being King of France (even granted that today France is a republic) is within the realm of the possible. For Graeber, political power is fundamentally linked to persuasion and the establishment of a belief. Thus, the political dimension is the 'dimension of social life in which things really do become true if enough people believe them'. But Graeber also underlines the limiting factor of this game of persuasion – like a magic trick, it only works if the players never openly recognize its essence; thus, Graeber likens the political process to magic.

The magic of the process of creating political power lies in its force to create signification where there was a prior void. Like businesses capitalizing on goods and services that were previously common goods or without charge, the making of political power is creating signification (and thereby occupies potential power space) where there is none or occupying spaces that have been semanticized by the status quo. Castoriadis recognizes the process of signification in the making of political power:

DOI: 10.4324/9780429288180-1

> The throne of the Lord of signification stands above the Lord of violence. [...]
>
> The mightiest army in the world will not protect you if it is not loyal to you – and the ultimate foundation of its loyalty is its imaginary belief in your imaginary legitimacy.[2]

Society is thus largely dominated by imaginary belief in imaginary legitimacy, and Castoriadis deduces that:

> ...one finds a dawning recognition of the fact that the source of the law is society itself, that we make our own laws—whence results the opening up of the possibility of challenging and putting into question the existing institution of society.[3]

Both Graeber and Castoriadis thus underline that the realm of the possible is largely dependent on the process of imagining power. Of course, the forces of gravity still exist, and in the same way that people can't breathe underwater, no one can stop a flying bullet shot by the riot police.

From a reverse perspective, preserving political power means restricting the forces of the imagination, making sure that the power of the status quo appears unsurmountable and keeps 'the forces of magic' under the lid. The belief that there is no alternative to the existing political and economic arrangements ensures that the riot police do not have to shoot the bullets because the legitimacy of the status quo is created by an exclusion of any possibility of change. Discouraged citizens rather stay on their couches and unload their anger and frustration in online forums or embark on a permanent journey into the oblivion of escapism and entertainment.

Graeber underlines that in the current market-driven societies the economic forces hinder the political imagination to flourish. The neoliberal paradigm, despite variations in the political setups (democracy, autocracy), has total domination of the political imaginary; he claims that 'the neoliberal project really has been stripped down to what was always its essence: not an economic project at all, but a political project, designed to devastate the imagination, and willing'.[4] If none can imagine an alternative then, what Graeber calls 'the war against the imagination',[5] is effectively won by the promoters of neoliberal economic arrangements. The often-quoted dictum that it is easier to imagine the end of the world than the end of capitalism is an obvious indictment of the crisis of imagination.

Imagination and Imaginary: The 'Self' in the 'Us'

Cornelius Castoriadis and Charles Taylor have both reflected on the dimension of imagination that legitimizes political power, the dimension that makes it possible to become the King (or Queen) of France or else create a society with different priorities than the current arrangements, and termed it 'social imaginaries'. Social imaginaries in Taylor's definition are 'the ways people imagine their social existence, how they fit together with others, how things go on between them and their fellows, the expectations that are normally met, and the deeper normative notions and images that underlie these expectations'.[6] Castoriadis describes them as 'norms, values, language, tools, procedures, and methods of dealing with things and doing things, and, of course, the individual itself both in general and in the particular type and form'.[7]

Whereas imagination is the individual dimension of perceiving and prefiguring the potential of the world, imaginaries are collective beliefs and their manifestations in the political sphere.

As both Taylor and Castoriadis underline, norms and values are not created in the sphere of individual imagination but in the social imaginaries. Social imaginaries are inherently also political imaginaries and thereby define the realm of the possible, the thinkable and the articulable. Taylor sees the imaginaries as larger forces to structure the perception of reality which are articulated not in theoretical terms but 'in images, stories, and legends'.[8] Stories and images are thus internalized and structure the imagination. New stories that are the product of the imagination are ineffective if they stay in the individual imagination and only have the potential for change if they enter the public realm. As Geoff Mulgan emphasizes 'shared imagination can create new shared realities'.[9] In the framework of political imaginaries, this would mean that new imaginaries can create new realities.

To come back to Graeber's statement on the war on imagination, the more precise description would thus be that the neoliberals have won the war on political imaginaries.

Imagination or, in the collective, imaginaries, and their possibility to transcend the existing framework can thus be judged in terms of their ontological priority. Is the individual (imagination) or the community (imaginary) the focal point?

Hannah Arendt judged the individual imagination in positive terms as a force to transcend its individual point: 'Private conditions condition us, imagination and reflection enable us to liberate ourselves from them and attain that relative impartiality that is the specific virtue of judgement'.[10]

Imagination thus enables a liberation from private conditions and thus we can transcend them. But are Arendt's observations still valid in a neoliberal society with the individual as its sole focus? Has the neoliberal imagination created a hermetically sealed individual imagination that does not allow the creation of political imaginaries? From the point of their ontological priority, if hope and imagination are privatized it leads to a cementation of the individual point of view.

Arendt warned that too strong a focus on private interests would destroy both the private and the public sphere. She emphasized that if 'the only thing people have in common is their private interests',[11] then as a consequence 'both the public and the private sphere are gone, the public because it has become a function of the private and the private because it has become the only common concern left'. The reduction to the private sphere as a sole concern leaves the individual in a state of isolation or, as Adam Curtis called it, in a trap. Curtis describes the contemporary mind-set as a world of individualism in which 'everyone is trapped within their own feelings, trapped within their own imaginations'.[12]

Although in Europe, the Covid crisis and the consequences of the war in Ukraine have brought a new focus on the public sphere in rhetorical dimensions, in economic terms the neoliberal focus on the private sphere is still more than a residue of pre-crisis times. Despite the change in public rhetoric, the trap of the individual imagination has been fortified by the lockdowns and the self-isolation of large amounts of the society. The (post) pandemic imagination is thus further away from a political imaginary that could challenge the status quo. Despite the fact that the collective dimension has been the rhetorical focal point in the crisis situation, the emphasis has been on the sacrifice of individual rights, wealth, and possibilities – not on the development of new imaginaries.

The ontological priority of the imagination has been thus the individual even if the laisser-faire of neoliberalism, in times of global pandemics, was tempered with severe, crisis-related, restrictions of individual liberties. For the future, the question is whether the imagination remains trapped in its individual cage, while the state apparatus and the economic elites continue to shape the parameters of societies, or if political imaginaries develop new forms of societal arrangements. Immanuel Wallerstein explicitly mapped out the two alternatives for the decades to come:

> If we are making a fundamental historical choice in the next fifty years, what is it between? Clearly, our choice is between a system (analogous to the present one in some fundamentals) in which some have significantly greater privileges than others, and one that is relatively democratic and egalitarian.[13]

Wallerstein stated this in 1998 when the economic ascension of autocratic economies, such as China, was not yet so evident and, obviously, Wallerstein had no premonition in regard to the impact of the pandemic on the development of Western democracies. But the alternatives at the point of bifurcation in the second decade of the 21st century still remain the same: focus on community or focus on the individual? Focus on political freedom or focus on economic freedom? Focus on new imaginaries or remain in the current trap of individual imagination?

In order to be a change agent on a larger scale, the ontological priority of the imagination thus has to be collective and transcend the individual point of view. Haiven and Khasnabish emphasize the communal aspect of the imagination:

> we understand the imagination as not merely the 'private property' of the individual. Through shared experiences, language, stories, ideas, art and theory we share part of our imagination. We create, with those around us, multiple, overlapping, contradictory, and coexistent imaginary landscapes, horizons of common possibility, and shared understanding.[14]

Even if Haiven and Khasnabish use the term 'imagination', what they are referring to are imaginaries. The focus on common possibility indicates that through a prism of private (property) imagination, shared horizons cannot be created.

Agency, Spatiotemporal Scope, and Emotion

A closer examination of the contemporary political imaginaries in the hybrid setting of neoliberalism and crisis-driven state control is thus not a mere intellectual exercise but a fundamental necessity in order to effectively address the shortcomings of the 'There Is No Alternative' doctrine that still pervades the public discourse and thus paves the way for neoliberal and state authoritarian policy making.

In the contemporary political context, the classic leftist/rightist division is obsolete. In an increasingly polarized public sphere none of the former political camps is offering political imaginaries that are transcending the neoliberal horizon. The political described by Chantal Mouffe as 'a vibrant "agonistic" public sphere of contestation where different hegemonic projects can be confronted',[15] has been reduced to a sphere in which the critique of the fundamentals of the economic hegemonic project is marginalized. Instead, as Mouffe claims, we 'are witnessing the unchallenged hegemony of neo-liberalism with its claim that there is no alternative to the existing order'.[16] In the absence of alternative hegemonic projects, the political discussion has shifted to moral categories with convenient good/evil categories that reduce the possibility of open discussion and enhance the antagonisms between the political groups.

A taxonomy of political imaginaries has to include the potential for sociopolitical change. Three criteria are henceforth suggested to assess their capacity to create new forms of societies, new stories, new images, and new values and norms: agency, spatiotemporal scope, and dominant emotional drive.

Agency

Any imaginary explicitly or implicitly defines the role of political agency, how much potential and scope for agency can be projected for the citizens, and how much is the agency delegated to others? In essence, the question of agency – the capacity of the political agent to act, to participate in the democratic debate – is about the potential of being able to effect change in the outcome (either through the articulation of his opinion, political activism, or through the establishment of alternative life models). In Mouffe's terms, political agency should have the potential to articulate and effect alternatives that challenge the hegemonic project of neoliberalism. Goodwin and Tyler (1981) call for a reconstructive political theory and name the three prerequisites:

1 a conception of society as an artefact (unlike Nature), capable of being purposefully altered by man himself
2 the conviction that progress, qua improvement, is possible
3 an analysis of socio-political life which is free from fatalism.[17]

Thus, political agency is possible if society is not understood as an immutable entity. Projecting an optimistic attitude to the possibility of change and an overall acknowledgement that activism can lead to change are essential factors in the process. This dynamic understanding of political agency and social change is based on the idea that every society is a social construction that can also be reconstructed by its members. So political agency and alternative hegemonic projects can be seen, according to Ruth Levitas:

> as a socially constructed response to an equally socially constructed gap between the needs and wants generated by a particular society and the satisfactions available to and distributed by it. All aspects of the scarcity gap are social constructs, including the propensity to imagine it away by some means or other.[18]

In brief, political agency in this context may be defined as the capacity to construct social responses to the socially constructed scarcity gap that have the potential to dynamize the debate on progress and challenge the hegemonic neoliberal project. As Hannah Arendt emphasizes 'men constantly create their own, self-made conditions'.[19]

Political imaginaries can thus be mapped according to their concept of political agency. The specific focus is on the question whether they articulate a concept of collective agency, since (wo)men can create their own, self-made conditions and find social constructs in relation to the scarcity gap in collective agreement.

Spatiotemporal Scope

Imaginaries that create new forms of societies, new stories, new images, and new values and norms have a temporal scope that they relate to. In 1516, with the publication of Utopia, the original title was 'De optimo rei publicae statu deque nova insula Utopia', Thomas More not only articulated an alternative political imaginary to an England that was dominated by enclosures, growing social inequality and corruption, but he also placed it in an ambigious space. The neologism Utopia can refer to the Ancient Greek 'eutopos', so good place, or 'outopos', so no place. More amplified the ambiguity with the choice of the name of the protagonist Hythloday, who discovers the ideal commonwealth on the island of Utopia; the name can be translated as 'skilled purveyor of nonsense'. Is Utopia thus a non-place or a better place or both? In any case, Hythloday is deeply impressed by the social arrangement on the 'nova insula Utopia'. In More's book, he gives an account of his findings:

> Thus have I described to you, as particularly as I could, the constitution of that commonwealth, which I do not only think the best in the world, but indeed the only commonwealth that truly deserves that name. In all other places it is visible, that while people talk of a commonwealth, every man only seeks his own wealth; but there, where no man has any property, all men zealously pursue the good of the public [...] for in other commonwealths, every man knows that unless he provides for himself, how flourishing soever the commonwealth may be, he must die of hunger; so he sees the necessity of preferring his own concerns to the public; but in Utopia, where every man has a right to everything, they all know that if care is taken to keep the public stores full, no private man can want anything; for among the area no unequal distribution, so that no man is poor, none in necessity.[20]

A new commonwealth with no property, no hunger, a concern for the public good and an equal distribution of wealth was as much of a utopian image (in the prerogative sense of the word) in the 16th century as it is today. But yet, why did More have to displace this ideal commonwealth, to locate it far away, if this was all just an account by a 'skilled purveyor of nonsense'? The answer is that Utopia became powerful as a political imaginary or, as J.O. Hertzler observed, before More's Utopia 'society was asleep'.[21] And Utopia animated others to articulate their political imaginaries: Tommaso Campanella wrote 'City of the Sun' (1623), Francis Bacon 'New Atlantis' (1627), and James Harrington 'The Commonwealth of Oceania' (1656). In the 18th and 19th centuries, many others would follow in their articulation of ideal societies. But all these ideal commonwealths were either displaced in time or in space. It was only the 19th century Utopian socialists that funded concrete local utopian projects and thereby put the 'no places' on the world map.

Whereas More's depiction of a good society through a strategy of spatial displacement provided a rich ground for the development of political imaginaries, in the contemporary context the question of spatial localization of political imaginaries is crucial. Unlike in the 16th century, 21st century citizens have more global awareness, more complex public spheres and a more critical attitude to overly idealistic accounts (due to a wider exposure to stories and historical narratives).

The temporal scope of political imaginaries is, ironically, also influenced by Thomas More but in a dialectically opposite way. More's Utopia, the good 'non-place', gave birth to the opposite – the Dystopia, the bad and hard place. The term (in the spelling of 'Dustopia') appeared for the first time in Lewis Henry Younge's *Utopia: or Apollo's Golden Days* in 1747. Whereas dystopias can have several functions (some of them anti-utopian, some of them a critique of the status quo, some of them a warning of harmful tendencies), most of them use temporal displacement as a narrative strategy.

In the current political context where the 'There Is No Alternative' (TINA) doctrine is dominating the discourse, any kind of spatial or temporal displacement of political imaginaries would lead to an immediate dismissal and a delegation of the ideas into the realm of fiction. So, in simple terms, any political imaginary has to answer the question: where does it happen and when does it happen? Or else, which concept of the ideal society do they, manifestly or covertly, articulate? Are they located in the 'here and now', the '(Western) idealized past', or the 'global apocalyptic future'?

Spatiotemporal scope thus refers to the question of localization of the prospective horizons of political imaginaries in the past, the present, and the future on a temporal level and to the abstract, the specific, and the global on a spatial level. It also involves the question of the normative approach to future images. Is the future perceived as positive, negative, or absent?

Dominant Emotional Drive

Political imaginaries share the essential features of good storytelling As Seargeant has observed: 'You pin your story around two emotions: hope or fear. This is what drives the desire for change; it's this which leads to the conflicts which give to shape the plot'.[22]

So, the basic emotional drives of political imaginaries are also defined by the two emotions: hope and fear. While Barack Obama's first presidential campaign famously was pinned around the concept of hope and successfully got him elected, more recent political leaders are centering their campaigns around fear. Greta Thunberg stated in her speech in Davos in 2019: 'I do not want you to be hopeful. […] I want you to feel the fear I feel every day. And then I want you to act'.[23] So how can the desire for change be driven more efficiently – by appealing to hope or to fear?

Matthew Barret and Mel Gilles argue that, despite the obvious need for action in relation to the dangers of climate change, the appeal to fear might actually prevent people from taking political action. They criticize the danger of an excessive use of fear images and messages and underline: 'The danger of the media's conflation of apocalyptic scenarios is that it leads us to believe that our existential threats come exclusively from events that are beyond our control'.[24] In their understanding, the apocalyptic storyline based on fear leads to either denial or despair at the feeling of inevitability.

In her book *The Politics of Fear*, Ruth Wodak, describes how fear is used by populist right-wing groups to whip up xenophobia, chauvinism, anti-intellectualism, and the scapegoating of others. Politics of fear in this context is designed to advocate strong leadership, exclusion, and ethnocentrism. So, fear closes people down, makes them more vulnerable, and potentially more aggressive towards people with different worldviews. As Edmund Burke wrote: 'No passion so effectually robs the mind of all its powers of acting and reasoning as fear'.[25] Fear thus induces passivity and irrational behavior.

Barry Glassner demonstrates how fearmongers deploy narrative techniques, such as repetition of wrong facts, making the irregular seem regular, and misdirection in order to create meta-narratives of fear. He describes how the fearmongering in the aftermath of 9/11 was used as a justification for wars and discrimination. Glassner concludes that fearmongers 'deploy narrative techniques to normalize what are actually errors in reasoning'.[26]

The appeal to the emotion of fear is thus an appeal to irrationality, an inhibitor for action, and an appeal to denial or despair. It remains questionable why Greta Thunberg wants to appeal to such negative emotions for the urgent and necessary fight against climate change.

In *The Principle of Hope*, Ernst Bloch advocated for a more positive approach to social change. Bloch wrote:

> Hope, superior to fear, is neither passive like the latter, nor locked into nothingness. The emotion of hope goes out of itself, makes people broad instead of confining them, cannot know nearly enough of what it is that makes them inwardly aimed, of what may be allied to them outwardly.[27]

The broadening of the horizons, the rejection of fatalism, and the potential for common visions and action are important conditions for the possibility of developing imaginaries which are not keeping people confined in their emotions of denial or despair.

Bloch's important contribution is to add to the cold stream of economic conditioning the warm stream of human agency. In his view, utopianism without an understanding of the economic structure is therefore as ineffectual for transcending the existing conditions as pure economic determinism. So economic determinism does not represent the totality of a complex reality. Reality as the dynamic understanding of possibilities is highlighted by Levitas (2005) who underlines that 'reality for Bloch must include the horizon of future possibilities – possibilities which are always plural and which are dependent on human agency for their actualisation'.[28]

Bloch argued that the way out of the 'close world that has already become', a world that he considered to be contemplatively idealistic or merely passively contemplative, would be to understand the future as a genuine open process. This kind of open future cannot be apprehended by mere contemplation but needs to be approached by 'thinking towards changing the world and informing the desire to change'.[29]

Catriona McKinnon underlines the self-fulfilling prophecy aspect of hope when she underlines that 'hope generates a disposition to act so as to make the realisation of hope's objective more probable whenever possible, all else being equal'.[30] Hope has thus an important psychological dimension which McKinnon sees in the practical commitment to the objective. This aspect is also emphasized by Philip Pettit when he describes hope as 'cognitive resolve'[31] in which through a strategic decision to act as if the hoped-for prospect was likely to be obtained. In times of growing cognitive dissonance and complexity, Pettit's strategic decision to cut the Gordian knot by acting in a more hope-oriented fashion is supported by Cheshire Calhoun when she advocates a broadly unreflective sense of the limitations of the future. Calhoun's 'phenomenological idea of the successful future'[32] is contrasting with the general perception of the potential apocalyptic future dominated by wars, global epidemics, and climate collapse. It would be easy to dismiss Pettit's and Calhoun's ideas as naïve or escapist; but from a psychological point of view, they are more likely to produce the mindset required to deal with the apocalyptic potential of the future. The deeper psychological aspect of interpersonal hope on motivation, rationality, and sociability is underscored by Catherine Rioux, who states that hope 'influences our motivation, the rationality of our actions, our other emotions, and the unfolding of our mental lives, in addition to being a distinctive way of relating to others'.[33]

Ruth Levitas stresses the importance of the concept of hope in relation to agency and its important function to combat fatalism:

> it is fatalism that is the key issue. Where it is no longer assumed that social organization is inherently controllable by human agents, or where it is no longer believed that the agents who are in control can themselves be made accountable to the rest of us, much of the motive for the construction of utopias as goals is lost. They cease to be images of hope-for future and become again expressions of desire.[34]

Cornel West warns about the oversimplification of the concept of hope. West calls for 'hope on a tightrope' and explains: 'real hope is grounded in a particularly messy struggle and it can be betrayed by naïve projections of a better future that ignore the necessity of doing the real work'.[35]

Of course, the activating potential of hope can also be reversed when the hope gets disappointed. The people who started to gather at Maidan Square in 2004, at the square in Zuccotti Park in 2011, at Tahir Square or Taksim Square in 2013 and were animated by the hope for social change experienced the painful aftermath. Their hopes were dashed, the squares returned back to normal life, and often the activists converted their hope into bitter cynicism and squared despair about the potential for change. Hope in this context did not provide a cognitive resolve or a successful future, and it rather left many participants locked into nothingness as opposed to liberating them. But is the squared despair in the aftermath of these squished attempts at revolution invalidating Bloch's idea about the activating potential of hope? Or, are they the result of a flawed perception of the context and the opposing forces, a lack of organizational skills and appropriate strategies, an absence of a coherent political project to challenge the status quo – a lack of common imaginaries?

The failed uprisings can be used as evidence for the uselessness of political struggle, as an invitation to plunge into squared despair or as justification for a fatalism that is driven by deep cynicism about the status quo; or they could be seen as a call to change the scope, the focus, the strategies, and the tactics of resistance.

Despite the failures of political activism, any kind of meaningful political story and the desire for change is motivated by hope, so the political imaginaries have to be pinned around the emotion of hope in order to be effective and deploy their full emancipatory potential. Hope thus has to be located in the future (again).

The Four Dominant Political Imaginaries

According to three criteria (concept of agency, spatiotemporal scope, and dominant emotional drive) four dominant contemporary political imaginaries can be identified: the neoliberal imaginary, the converted imaginary, the apocalyptic imaginary, and the nostalgic imaginary. In a book chapter I wrote in 2019, I attempted to map the mind-space of political imagination,[36] and the apocalyptic and the nostalgic imaginary were both classified as 'apocalyptic imagination'. After the outbreak of Covid 19 and the Ukraine War, the media coverage, the policy responses, and the relatively obedient reaction of many people towards restrictive measures to combat the spread of the virus started to indicate a shift. Future research will hopefully create a clearer picture, but even without judging the measures taken in each country, it became clear that apocalyptic imagination is a very effective reference point for new policies that were unimaginable prior to Covid 19. In the book chapter, I reduced the scope of apocalyptic imagination by merely associating it to the nostalgic imaginaries of alt-Right groups and right-wing populists. The spread of apocalyptic imaginaries on a mass scale across all political spectrums during the Covid 19 epidemic indicated that it is fundamentally different from nostalgic imaginaries. Consequently, both imaginaries require a separate examination. In the overall, there are four archetypes in the taxonomy of political imagination.

The Neoliberal Imaginary

The core of the neoliberal imaginary has been articulated by the economist Milton Friedman: 'A society that puts equality before freedom will get neither.

A society that puts freedom before equality will get a high degree of both'.[37] Friedman defines freedom as 'whatever it is that interests the participants, whatever they value, whatever goals they pursue'.[38] Essentially, the freedom Friedman refers to is economic freedom. The as a societal choice to limit the freedom in favor of equality (and thereby an act of political freedom to construct social responses to the socially constructed scarcity gap) is implicitly excluded. Friedman thus excludes collective agency from the concept of freedom and puts the full emphasis on individual agency (to pursue 'whatever goals they pursue'). The TINA doctrine is a manifestation that alternatives to the existing politico–economic arrangements are, per definition, inferior to the status quo. In the absence of a political project that supersedes the free markets, the individual is left in an atomized struggle to fulfill his privatized imagination. The neoliberal imagery excludes or marginalizes other imaginaries and could thus be qualified as an anti-imaginary. The concept of agency is reduced to an individual scope and thereby effectively neutralized.

The spatiotemporal scope of neoliberalism is both wide and narrow. The spatial scope is wide in the sense that it created the foundations of the globalized economy. Global markets are the ultimate realization of Friedman's ideas of economic freedom (which privileges those with the biggest amount of capital). The temporal scope, on the other hand, is more constrained. Progress is measured in numbers of economic growth, the focus is on the here and now. Future images project an idea of the society which is based on 'more of the same'. More economic growth, more wealth, a bigger expansion of the neoliberal system and the global markets and, of course, more economic freedom. The aim is to maintain the neoliberal approach to maintain the intersubjective realities which is meant to ensure the further growth of global wealth. The normative dimension is absent. In addition, the neoliberal imaginary is not projecting any collective future images.

The emotional drive is based on hope, but the hope is strictly limited to the private domain and limited to those who are financially able to compete for the increasingly scarce resources. The hope is to be on the winning side of those who have access to the limited realm of economic freedom. In a generalized laisser faire, the privileges of the privileged are growing while the majority of 'the 99%' is betting on (at least in the West) the increasingly difficult possibility of upward mobility.

The Converted Imaginary

Since the end of the 19th century, an imaginary that opposed liberalism and championed alternative visions was on the rise: the socialist imaginary, be it in its communist or social-democratic version. Even if the fall of the Soviet Union was not the end of history as Francis Fukuyama claimed, the shockwaves of the implosion of Communism went through socialist and social-democratic movements all over the world. In the 1990s, social democratic movements started to adjust to the new realities, the self-perceived

New Left split off from the Old Left, largely abandoning the foundations of the economic programme of Social Democracy and converting to a more market-driven approach. As Anthony Giddens points out:

> A hundred and fifty years ago Marx wrote that 'a spectre is haunting Europe' – the spectre of socialism or communism. This remains true, but for different reasons from those Marx had in mind. Socialism and communism have passed away, yet they remain to haunt us. We cannot just put aside the values and ideals that drove them, for some remain intrinsic to the good life that it is the point of social and economic development to create. The challenge is to make these values count where the economic programme of socialism has become discredited.[39]

The departure from traditional social-democratic policies based on a strong state ensuring a well-financed public infrastructure, welfare systems, employment policies based on high salaries and stable jobs, equality in opportunities and outcomes and high taxation for the better off led to an (initially) well-concealed turn towards neoliberal policies. In 1998, in a common paper, the most prominent advocates of what was labelled the 'Third Way', British Prime Minister Tony Blair and German chancellor Gerhard Schröder were pressed to underline that 'Modern social democrats are not laisser-faire neo-liberals'.[40] But the policies introduced by Blair and Schröder, labor market deregulation, tax cuts for businesses, partial privatization of public sector activities, promotion of free-trade agreements, welfare cuts, indicated a conversion to the neoliberal imaginary. It is hard to articulate how these policies render 'the values and ideals' that drove socialism and communism, as Giddens underlines. The Third Way was thus largely a rhetorical figure for neoliberalism from the Left. Tony Blair underlined in many speeches that the New Labour policy aims at the freedom for many, echoing Friedman's primary focus of neoliberalism. Freedom for many is not equality for all. After eight years in office, in 2005, Blair articulated his neoliberal credo openly: 'In the era of rapid globalisation, there is no mystery about what works: an open, liberal economy, prepared constantly to change, to remain competitive'.[41]

In this New Left open, competitive, liberal economy which ensures the freedom for many, the focus on political agency is identical to the neoliberal imaginary. Anthony Giddens describes this shift as a reaction to the new realities:

> The 'me' generation is a misleading description of the new individualism, which does not signal a process of moral decay. Rather to the contrary, surveys show that younger generations today are sensitized to a greater range of moral concerns than previous generations were. They do not, however, relate these values to tradition, or accept traditional forms of authority as legislating on questions of lifestyle.[42]

Giddens underlines that the consequence to this shift, Third Way politics, abandoned the idea of collectivism. The new emphasis is thus on freedom, private morality, and lifestyle. This seismic shift away from a collectivist focus on economic equality dominates most social-democratic movements, even after the departure of Tony Blair and Gerhard Schröder from office. The strategic move towards the new centre or 'die neue Mitte' which proved successful in electoral terms has led towards the acknowledgement that agency was to be conceived in individual terms. The New Centre, the New Radical, the Progressive Neoliberal or the Life-Style Leftist have ceased challenging the hegemonic neoliberal project and instead converted to it and shifted the emancipatory struggle to lifestyle questions that can easily be co-opted by the neoliberal credo of choice.

As for the so called 'Old Left', there are occasional reminders that it still might have the potential to rally citizens around a collective project that aims at rekindling the economic foundations – such as the primary campaigns by Bernie Sanders, but on a global scale. The Post-Third Way is dominating social democratic policies. Even the state-interventionist policies during the pandemic have not challenged the fundaments of the new market-driven leftists.

Consequently, the mainstream leftist imaginaries have also adopted the spatiotemporal scope of neoliberalism. The spatial scope, as articulated by Tony Blair, is the global economy in which each nation state competes for investment and business opportunities. The traditional spatial scope of social democracy, the nation state, has thus shifted from a state providing equality and social welfare to a state that has to deregulate the labour market, downsize the public sector, and provide tax incentives to global businesses in order to remain competitive.

In the same vein, the temporal scope is defined by the push for economic growth which is the sole measure of progress in the new social democratic understanding. The converted imaginary also renounces projecting any future images that are other than 'more of the same'. Converted imaginaries have fully internalized Friedman's warning that equality should never be prioritized over (economic) freedom.

The former motivating drive of hope for a more equal society, for more social justice, for more opportunities, and more upward mobility are all based on challenging the economic foundations of (neo-)liberalism. Now that these aims have been abandoned, the hopes for a better society are more fragmented, more atomized and non-inclusive and more symbolic.

Even if there is no doubt that the fight for gender equality, against racial discrimination, and for more LGBTQ rights are of primordial importance because they represent an important part of the global fight for justice, as long as the economic basis for the discrimination through poverty remains the same, these emancipatory movements represent only a fragment of what social democratic movements used to fight for. As long as millions of single mothers have to work at two low-income jobs in order to feed their children, another academic post in gender studies does not create more equality. Of course, there is no point to pitting the symbolic battles against the economic battles, but as long as the emancipatory struggle is fragmented and largely focused on moral and symbolic issues, the converted imaginary will firmly remain in the realm of neoliberalism.

The Apocalyptic Imaginary

The Covid pandemic produced images of ambulance cars invading a virus-infected Bergamo, body-bags of Covid victims piling up next to the Brooklyn Bridge, and major metropoles in a complete standstill with empty streets and boulevards, in addition to leading to the creation of historically unprecedented global lockdowns that confined billions within their own four walls. Prior to the pandemic, such kind of events were the material for a Hollywood disaster film or the fantasies of obscure apocalyptic cults. While Covid was not the apocalypse, it has fueled apocalyptic imaginaries by providing 'end of the world' like events. As Simon Dein points out:

> Pandemics indicate the fragility of life and the world, chaos, engender paralysing anxiety that the world is dissolving, a sense of detachment and raise significant issues of meaning resulting in existential crises. Self-isolation and quarantine create a sense of being separated from the community and world generally—a sense of anomie.[43]

In this sense of fragility, anomie, anxiety and existential crises, Hollywood disaster film scenarios become perceivable, and apocalyptic visions and projections – not only those in relation to the Covid pandemic – acquire the aura of plausibility.

The Greek word 'apokalypsis' means revelation, and the story of the apocalypse is told in the Book of Revelations of the New Testament, the final book of the bible in which John describes his unveiling of the truth. One important element in the Book of Revelations is that believers are saved in the rapture, while the sinners stay on earth and face tribulation in which they have to suffer for their sins. In the end, Jesus will win the battle against the Anti-Christ and the believers will enter the New Jerusalem. The last chapter of the bible thus reveals ultimate destruction and suffering (for the sinners), a fierce battle between good and evil, and the resurrection and the New Jerusalem for the believers. The wisdom acquired in this apokalypsis is thus that the catastrophe has a cathartic effect. As John J. Collins points out, the logic behind the apocalypse is based on 'wisdom [...] not the inductive kind [...] but acquired through revelation'.[44] The truth has thus always been present but is unveiled in the moment of catastrophe. The implication is the ultimate truth is already present, waiting to be discovered through wisdom.

Religious apocalypticism had a vast influence on Christian thought. In the age of modernity, it was entailed by a secular version of apocalypticism. In secular apocalypticism, the future is not defined by a divine intervention, the end of the world comes through (an often man-made) catastrophe, such as atomic wars, flood catastrophes, melting glaciers, or global pandemics. In many of the secular apocalypses, the tribulation, the period of intense suffering and the battle between the Christ and the Anti-Christ are conflated in one cataclysmic event. Due to the lack of divine presence, the hope for the New Jerusalem becomes inarticulable. The secular apocalypticism thus downsizes or eliminates the element of hope articulated in the Book of Revelations. Essentially, doom is highlighted, and the idea of the New Jersusalem is also destroyed along with the catastrophe. Despite their secular orientation, these forms of apocalyptic thinking still contain the moral dichotomies between good and evil and also the claim of wisdom to unveil the absolute truth.

Barett Gross and Gilles link the secular apocalypticism to the loss of faith in the idea of progress and the deterioration of old-world views based on the unsustainability of the contemporary way of life in the face of the climate catastrophe. They maintain that 'the further we retreat into that apocalyptic fantasy, the more likely we are to bring about the apocalypse we fear'.[45] The expectation of a cataclysmic, crystallizing moment thus creates a moment of paralyzed anticipation.

Michael Shermer underlines that the human brain is a pattern-seeking belief engine and sees a therapeutic element in the conceptualization of the end of the world. As Shermer observes not only do they feed the pattern-seeking brain, but apocalyptic visons:

> help us make sense of an often seemingly senseless world. In the face of confusion and annihilation we need restitution and reassurance. We want to feel that no matter how chaotic, oppressive or evil the world is, all will be made right in the end.[46]

Matthias Riedl observes the same symbolic function but draws the attention to a more antagonistic element of the need for restitution:

> The apocalyptic visionaries diagnosed that the world was so totally corrupted that any engagement in politics or resistance had become meaningless; only a transformation of the entirety of reality would ultimately bring bliss to the believers and ruin to their enemies.[47]

In Riedl's interpretation, the apocalyptic imaginaries have thus an activating potential but the source of the desire to not act is not hope but alienation and resentment. Riedl notes that apocalyptic sentiments imply the scenario of a war of good against evil. The psychological aspect is the transformation of narratives of decline and the transformation of 'determinist quietism'[48] into activism. They are thus, as Riedl emphasizes, symbolic expressions of proleptic existence, 'an existential attitude where the alienation from present reality is almost complete and all meaning of life rests on the anticipation (Greek: *prolepsis*) of a future transformed reality'.[49]

Apocalyptic imaginaries are thus based on an activating potential, but they are based on ontological precarity, alienation, polarization, the claim for absolute wisdom, and the absence of a hope image. In this context, Greta Thunberg's 2019 Davos speech with its emphasis on fear and the ensuing appeal to action appears to be a divisive statement and not a call for humanity to stand up to the menace of climate change.

The explicit reference to apocalyptic imaginaries is manifested in the names of some of the radical climate action movements: Extinction Rebellion or Letzte Generation (Last Generation). While there is no question of the necessity to combat climate change and the need for activism, the apocalyptic imaginaries used by these movements are based on fear and polarization. Whether these are the right emotions to affect change on a global scale remains questionable.

Apocalyptic imaginaries are constituting elements of many political movements whether they are Islamic fundamentalists, right-wing terrorists, or anti-vaxxers, and they also have pervaded the mainstream media discourse on the Covid pandemic. Even mainstream political parties implicitly or explicitly appeal to apocalyptic imaginaries whether they are based on images of mass migration, economic meltdowns, or climate catastrophe. An important element in the apocalyptic imaginary is the appeal to crisis.

The etymological root of crisis is the old Greek 'krisis' which means 'decision'. So, the 'revelation' of the apocalypse is that there is a crisis in which a decision has to be taken (in the Book of Revelation, the decision was to accept Jesus as a savior and thereby be part of the rapture). The apocalyptic imaginaries do not articulate any clear agency about who should take the decision (often the undefined 'we' stands as a placeholder). Despite the fierce criticism of the status quo and the appeal to collective action, they are not fundamentally challenging the neoliberal status quo. As opposed to being beaten down by the police during their demonstrations or strikes, the climate change activists, for instance, are invited to speak in Davos. The less influential political movements that are shaped on apocalyptic imaginaries are often communities of opinion on the internet with no real-world activating power. In the face of co-optation of the climate change movement, the justified fear of climate change can be a powerful tool for political decision-making. In the absence of a clear concept of agency (and concrete ideas) that challenges the status quo, the crisis allows for political change that fuels further neoliberal arrangements. Milton Friedman asserts the potential of the crisis for political action:

> Only a crisis – actual or perceived – produces real change. When that crisis occurs, the actions that are taken depend on the ideas that are lying around. That, I believe, is our basic function: to develop alternatives to

> existing policies, to keep them alive and available until the politically impossible becomes the politically inevitable.[50]

The focus on fear and the endpoint of total transformation allows only defensive reactions in the face of crisis, thus the concept of agency in the apocalyptic imaginaries impairs the construction of social responses to the socially constructed scarcity gap.

The spatiotemporal scope of the apocalyptic imaginaries makes immediate action or even the prospect of effective change difficult to achieve. The spatial scope (the end of the world) is per definition global, so even if local actions might have an effect, they can be potentially rendered completely useless if in other world regions the situation is aggravated tenfold.

As for the temporal scope, the horizon expectation is fixed to the future, whether nearer or further, with an element of epistemological precarity attached as to when the future comes into being (with the exception of sectarian movements that predict the end of the world on a precise date).

The dominant emotional drive for the secular apocalyptic imaginaries is fear, not only in relation to the present but even more in relation to the future, and the fear factor is set to increase with the deterioration of the global conditions.

The Nostalgic Imaginary

The sentiment of longing for a period in the past that always had a profound influence on the human condition. In 'Requiem for a Nun', William Faulkner famously wrote: 'The past is never dead. It's not even past'.[51] In political imaginaries the past has always constituted an important reference point in order to legitimize power. Arguably, the 21st century political discourse in the Western world has boosted references to the past and appeals to feelings of nostalgia to new levels. The demographic change in Europe that led to aging societies, the decline of geopolitical and economic significance in a multipolar world, and the speed of technological change are factors that have contributed to longing for a (imaginary) 'golden age' which is located in the past. The 21st century nostalgia has its root in the previous century. Svetlana Boym asserts that:

> The twentieth century began with utopia and ended with nostalgia. Optimistic belief in the future became outmoded, while nostalgia, for better or worse, never went out of fashion, remaining uncannily contemporary.[52]

The absence of hope for the future has thus led to a growing reference to the past and a cultural malaise in relation to the future. In addition, the fact that in most European societies the majority of the electorate is above 40 and thereby biologically more prone to see their past (youth) as a more optimistic place has shifted the political imaginaries.

Right-wing populist movements, such as the Rassemblement National in France, the Fratelli d'Italia in Italy, the Alternative für Deutschland in Germany, or the Vox in Spain are capitalizing on the feeling of nostalgia. Nostalgia translates as 'longing for a return home' (from the Greek 'nostos' and 'algia'); so, as an emotion, it is perfectly suited to painting the home of the past in golden colors and using them as fantasies of restoration. But, of course, the 'home' depicted by the right-wing populist is used as a metaphor of exclusion and as a tool to discriminate against those who are not part of the ethnic community or who are not conforming to traditional ideas of gender or sexual orientation. The nostalgic imaginary of right-wing populism is based on distorted images of past vitality and thus a return to a home that never existed. In the American context, Naughton and Vlasic observed that nostalgia 'is not simply about America reliving a Golden Age. It is about reinterpreting it'.[53] The return to home of right-wing populism is thus also about refurbishing the past for ideological purposes.

Svetlana Boym distinguishes between two types of nostalgia, the restorative and the reflective. While the restorative nostalgia puts the emphasis on nostos (home) and attempts a transhistorical reconstruction of that home, the reflective nostalgia is focused on the algia (the longing itself) and is thus a more diffuse and personal emotion. Right-wing populism is constructed on restorative nostalgias with a rhetorical focus on what are considered to be community-building values (family, nature, homeland). But, as Boym warns, 'restorative nostalgia returns and rebuilds one homeland with paranoic determination'.[54] In addition, the restorative nostalgia camouflages the fact that it is a nostalgia and instead puts claims on truth and tradition. Boym asserts:

> *Algia* (or longing) is what we share, yet *nostos* (or the return home) is what divides us. The promise to rebuild the ideal home lies at the core of many powerful ideologies today, tempting us to relinquish critical thinking for emotional bonding.[55]

Boym claims that the beginning of the 21st century is not characterized by a search for newness but by the proliferation of nostalgias which includes an emphasis on traditions that are often very recent and based on a selective approach to the past. Thus progress, unlike that projected by 19th-century enlightenment thinkers, is not eliminating nostalgia but rather exacerbating it. The longing for a return to home, as Boym explains is thus not 'a longing for a place, but it is actually a yearning for a different time'.[56]

Fred Davis divides nostalgias into private domain nostalgias and public domain nostalgias. The focus in the nostalgic imaginary is on the usage in the public domain. But private emotions such as a yearning for a different time, especially when they simultaneously appear on a mass scale, equally have a strong impact on the political process. Davis sees nostalgia as a continuing quest for personal identity and emphasizes the comforting aspect of nostalgias: 'In the clash of continuities and discontinuities with which life confronts us, nostalgia clearly attends more to the pleas for continuity, to the comforts of sameness and to the consolations of piety'.[57] Davis argues that nostalgia 'always occurs in the context of present fears, discontents, anxieties, or uncertainties'.[58]

In the same way as the apocalyptic imaginary, the nostalgic imaginary is thus a reflection of an alienation from the present. But as opposed to projecting apocalyptic visions into the future, the nostalgic imaginaries which are based on restorative nostalgias are attempting to react to the discontinuities of the present by projecting the past as a golden age. The escape from the present time, from the beginning of the 21st century, is the common motivation for both imaginaries. Faced with the overwhelming desire to escape, the apocalyptic imaginary programmes the time machine in a dystopian future and the nostalgic imaginary takes the opposite temporal direction. As Zygmunt Bauman has pointed out, a utopia that, as opposed to being projected in the future, is constructed on a dead past is a 'retrotopia'.[59] These retrotopias can have a strong activating potential without offering a vision of the future (other than returning to the past).

The power of these nostalgic imaginaries has been described by Mark Lilla who claims that hope can be disappointed, but: 'Nostalgia is irrefutable'.[60] The irrefutability of restorative nostalgias is thus the engine that drives the reactionary right-wing populist movements and enables them to argue in the post-factual realm and to emotionalize the political discourse. This also creates a strong attachment to their leaders which cannot be undermined by any rational discourse. The 'home' of right-wing populists is a castle that is meant to protect against the complexity, diversity, discontinuity, and anxieties of the present. It also serves as a fortress against the fear(s) of the future.

The concept of agency in the nostalgic imaginary is limited to the delegation of political power to the leader. The leader persona implicitly or explicitly promises a path of return from the alienation of the present to the idealized home of the past with the aim to eliminate the discontinuities and anxieties of the future. The collective agency is animated by the instrumentalization of private longings and incorporated by a 'savior'.

The spatial scope of this imaginary is the home, the region; the nation state is its temporal scope: the (reinterpreted and reconfigured) past. The future image is a sepia-tinted, monochrome portrayal of a society where anxieties, instabilities, and complexities did not exist, a controlled fantasy where the world remains unchanged and intelligible. Boym describes this desire for (restorative) nostalgia as 'a mourning for the impossibility of mythical return, for the loss of an "enchanted world" with clear borders and values'.[61]

The dominant emotional drive is thus fear of the future alimented by the hope for a return to the past (home). In its backward orientation, the nostalgic imaginary thus does not fundamentally challenge the neoliberal hegemony. Often their leaders are using the nostalgic imaginaries as a smoke screen to conceal the continuation of the same economic policies that have led to instability and discontinuity in their voter's life. The nostalgic imaginary is thus dangerous for its inherent scapegoating, xenophobia, homophobia, and misogyny. The channeling of negative emotions to the sphere of 'values' disguises the fact that the nostalgic imaginary is in fact in its politico-economic foundation what Nancy Fraser calls '*reactionary neoliberalism*'.[62] The effect of this reactionary liberalism and its media culture focused on the short term is the absence of new ideas. As Mark Fisher explains this paradox:

> On the one hand, this is a culture that privileges only the present and the immediate – the extirpation of the long term extends backwards as well as forwards in time (for example, media stories monopolize attention for a week or so then are instantly forgotten); on the other hand, it is a culture that is excessively nostalgic, given over to retrospection, incapable of generating any authentic novelty.[63]

The Imaginary Mind Scape and Its Impact on Agency

Neoliberal, converted, apocalyptic, and nostalgic imaginaries are dominating the mind scape of the 21st century. To come back to the short definition of political agency as the capacity to construct social responses to a socially constructed scarcity gap, none of the four imaginaries are providing a concept of agency that is both collective and oriented towards the future. While the neoliberal and the converted imaginaries have an orientation towards the present but (ever since the rejection of social democratic movements to privilege equality over economic freedom) they are based on individual agency and a lack of collective future images. The collective aspect is emphasized (in an abstract 'we') in the apocalyptic and nostalgic imaginaries, but both imaginaries are displaced in time, either projecting dystopian fear into the future or escaping into a glossy falsification of the past.

The spatial scope of the neoliberal and the apocalyptic imaginaries are global. But their perspective is profoundly based on concepts of either Western (Chicago School) economics or (secularized) Christian perspectives of the end of the world. The converted and the nostalgic imaginaries are predominantly focused on the nation state, either, in the converted perception as a competitor on the global market, or, in the nostalgic version as the source of meaning and the sole provider of continuity.

The dominant emotional drive of the neoliberal and converted imaginaries is the privatized hope. So, hope has been disconnected from the society and moved into the realm of the private and thereby neutralized as a motivator for social change. Apocalyptic and nostalgic imaginaries are based on fear and articulations of a profound malaise with the present and an intense pessimism in regard to the future. Both have an authoritarian impulse that reflects the desire for stability in a context of an individually perceived lack of control and continuity.

The neoliberal imaginary, despite the growth of control societies in the context of the pandemic, is not challenged in its hegemony by either the converted, the apocalyptic, or the nostalgic imaginaries. These three imaginaries have renounced from challenging the core of the economic system and articulate their political visions largely in the quest for specific reforms that are not antagonistic. Thus, strong border control regimes articulated by right-wing populists are not in conflict with the continuation of unequal trade relations with the Global South; the demands of the climate change activists are driving new potentials of economic growth in some sectors of the economy; and the social democratic call for unconditional basic income can potentially become reality in an efficiency-driven artificial intelligence (AI) society.

None of these imaginaries expands the boundaries of the possible, the thinkable, and the articulable when it comes to the redefinition of political power, the establishment of new forms of organizing the scarcity gap, or a reimagination of political agency.

Frederic Jameson observed already in the 1990s that there was a paradox in the unparalleled rate of change on all levels of social life and an unparalleled standardization of everything. Jameson wrote: 'What then dawns is the realization that no society has ever been so standardized as this one, and that the stream of human, social, and historical temporality has never flowed quite so homogeneously'.[64] The reduction of the scope of the contemporary imaginaries and the refusal to challenge the neoliberal hegemony is exacerbating the standardization and the homogenization of political thought. The climate crisis, the global pandemic, and the transition from a unipolar to a multipolar world order (and the increasing potential for war and military conflicts) are making the absence of new political imaginaries even more acutely felt. As George Monbiot points out: '...order in times of scarcity is not the same as order in times of plenty. It is harsher and less flexible: the realities of power are more keenly felt'.[65] Zygmunt Bauman also articulates the growing pressure: 'From the top to the bottom of society [...] a climate of mutual (and a priori) mistrust, suspiciousness, and cut-throat competition is created. In such a climate the germs of communal spirit and mutual help suffocate, wilt, and fade'.[66]

If none of the four imaginaries described above can challenge the hegemony, then it is time to reconsider the parameters for an imaginary with the potential to create new political possibilities. The fantasies of revolution, cosmopolitan citizenship, and transformation through new symbolic representation should be delegated to the advertisement world. A new political imaginary should not repeat the mistakes of the 20th century and fulfill the delusions of fanatics (or justify the passivity of those waiting for the revolution) and therefore put the idea of violent uprising to the dustbin of history. The reverie of a world citizen in an increasingly conflictual multipolar world is a beautiful ideal but not a practical orientation for political action. And the lifestyle leftist obsession with moral attitudes and identity politics is obscuring the economic roots of the current disorder.

As long as the possibility of change is in the abstract and activism (in the vein of the squared revolutions) leads to disillusion and retreat into privatized cynicism, the potential for political agency is downsized and atomized. Without the articulation of images, stories, values, and norms that reflect shared imaginaries and project the possibility of shared existence, the potential for social change remains in the realm of the four imaginaries that are not challenging the hegemonic order.

Since the global scope is too large for effective change, and the state is too intertwined with the global system, then the spatial scope has to be downsized in order to upsize the potential for political action. Bauman explains that as a result of globalization 'states are presently turning into not much more than somewhat larger neighbourhoods [...] while the neighbourhoods of old struggle to take on the role of little states'.[67]

In order to open the horizons, the imaginaries of change have to be plural and activating and thereby driven by hope. Hope, as a cognitive resolve and the potential for self-fulfilling prophecies, creates horizons of expectations which, in the scope of 'neighbourhoods', can be fulfilled more realistically than on an abstract global scale. But Cornel West's warning that hope fulfillment is an ongoing messy struggle is a useful reminder that naïve hope fulfillment and the focus on the short term are potential pitfalls. Ironically, the most iconic neoliberalist, Milton Friedman, demonstrated that a fringe and minority opinion within the field of economy can become the new hegemony if a strong belief in ideas is combined with patience and resilience. It took roughly 20 years before his ideas were implemented on a state level (in Chile in the mid 1970s) and another 20 years before they were gradually introduced on a global level (after the breakdown of the Soviet Union in the early 1990s). In 1951, as a young economist, Friedman commented on the power of faith in ideas:

> Ideas have little chance of making much headway against a strong tide; their opportunity comes when the tide has ceased running strong but has not yet turned. This is, if I am right, such a time, and it affords a rare opportunity to those of us who believe in liberalism to affect the new direction the tide takes. We have a new faith to offer; it behooves us to make it clear to one and all what that faith is.[68]

Many activists that are not activated by such a strong hope in the prospect that 'the politically impossible becomes the politically inevitable' give up their (often strategically misguided) struggle to develop alternatives after just a few weeks, months, or years. Obviously, this does not diminish their contribution, for activism is always better than passivism, but it delineates the potential temporal scope of political struggles. Other than Milton Friedman's ideas, political imaginaries based on the local level do not (initially) have a global scope, and the potential to see positive results in a shorter time period than 20 years is immanently realistic. Still, political change is a messy business with little more as a security than the faith in strong ideas and in good strategies to implement them. And once they are established, the (imaginary) legitimacy has to be established. As Immanuel Wallerstein underlines: 'Re-creating some kind of social order is a matter not only of constructing an alternate system but also, in very large part, of legitimating that constructed system'.[69]

But, the starting point is to engage with ideas that have the potential to create spaces of possibility and new imaginaries.

Notes

1 David Graeber, *Revolutions in Reverse. Essays on Politics, Violence, Art and Imagination* (New York: Minor Compositions, 2011), p. 94.
2 Cornelius Castoriadis, *Philosophy, Politics, Autonomy. Essays in Political Philosophy* (Oxford: Oxford University Press, 1991), pp. 143–175 (pp. 155–156).
3 Cornelius Castoriadis, *World in Fragments: Writings in Politics, Society and the Imagination* (Stanford: Stanford University Press, 1997), p. 86.
4 Graeber, pp. 5–6.
5 Graeber, pp. 112–113.
6 Charles Taylor, *Secular Age* (Cambridge, MA: Harvard University Press, 2007), p. 171.
7 Castoriadis (1997), pp. 6–7.
8 Charles Taylor, *Modern Social Imaginaries* (Durham: Duke University Press, 2004), p. 23.
9 Geoff Mulgan, 'The Imaginary Crisis (and How We Might Quicken Social and Public Imagination)', *UCL Demos Helsinki and Untitled* (2020), 29 <https://demoshelsinki.fi/julkaisut/the-imaginary-crisis-and-how-we-might-quicken-social-and-public-imagination/> [accessed 23 April 2022]
10 Hannah Arendt, *Lectures on Kant's Political Philosophy*. Edited and with an Interpretive Essay by Ronald Beiner. (Brighton: Harvester, 1982), p. 73.
11 Hannah Arendt, *The Human Conditions* (2nd ed.). (Chicago: The University of Chicago Press, 1958), p. 69.
12 Adam Curtis, as cited in Mark Fisher, *Capitalist Realism: Is There No Alternative?* (Winchester, UK: O Books, 2009), p. 74.
13 Immanuel Wallerstein, *Utopistics. Or, Historical Choices of the Twenty-First Century* (New York: The New Press, 1998), p. 69.
14 Max Haiven and Alex Khasnabish, *The Radical Imagination: Social Movement Research in the Age of Austerity* (Halifax: Fernwood Publishing and London: Zed Books, 2014), p. 3.
15 Chantal Mouffe, *On the Political* (London: Routledge, 2005), p. 3.
16 Ibid., p. 33.
17 Barbara Goodwin and Keith Taylor, *The Politics of Utopia: A Study in Theory and Practice* (London: Hutchinson & Co, 1981), pp. 23–24.
18 Ruth Levitas, *The Concept of Utopia* (Witney: Peter Lang, 2011), p. 210.
19 Arendt (1958), p. 5.
20 Thomas More, *Utopia* (New York: Dover Publications, 1997), p. 81.
21 Joyce O. Herzler, *The History of Utopian Thought* (London: George Allen & Unwin, 2012, original work published 1923), p. 39.

22 Philip Seargeant, *The Art of Political Storytelling* (London: Bloomsbury Academic, 2020), p. 127.
23 Jonathan Watts, 'Interview: Greta Thunberg, Schoolgirl Climate Change Warrior: "Some People can Let Things Go. I Can't"', *The Guardian*, March (2019) <https://www.theguardian.com/world/2019/mar/11/greta-thunberg-schoolgirl-climate-change-warrior-some-people-can-let-things-go-i-cant> [accessed 20 January 2021]
24 Matthew Barret and Mel Gilles, 'How Apocalyptic Thinking Prevents Us from Taking Political Action', *The Atlantic*, April 23 (2012).
25 Edmund Burke, *Philosophical Enquiry into the Origin of Our Ideas of the Sublime and Beautiful* (London: R. and J. Dodsley, 1757), p. 42.
26 Barry Glassner, *The Culture of Fear: Why Americans Are Afraid of the Wrong Things: Crime, Drugs, Minorities, Teen Moms, Killer Kids, Mutant Microbes, Plane Crashes, Road Rage, & So Much More* (New York: Basic Books, 2001), p. 829.
27 Ernst Bloch, *The Principle of Hope* (reprint ed. 1995), trans. by N. Plaice, S. Plaice, and P. Knight (Cambridge, MA: MIT Press, Vol. 3:3, 1986), p. 15.
28 Ruth Levitas, *The Concept of Utopia* (Witney: Peter Lang, 2005), p. 7.
29 Bloch, p. 8.
30 Catriona McKinnon, 'Cosmopolitan Hope', in *The Political Philosophy of Cosmopolitanism*, ed. by G. Brock and H. Brighouse (Cambridge: Cambridge University Press, 2005), pp. 243–249 (p. 245).
31 Philip Pettit, 'Hope and It's Place in the Mind', *The Annals of the American Academy of Political and Social Science*, 592 (2004), pp. 152–165.
32 Cheshire Calhoun, *Motivating Hope. In Doing Valuable Time: The Present, the Future, and Meaningful Living* (Oxford: Oxford University Press, 2018), pp. 68–90.
33 Catherine Rioux, 'Hope: Conceptual and Normative Issues', *Philosophy Compass*, 16(3) (2021), p. 7.
34 Levitas (2011), p. 226.
35 Cornel West, *Hope on a Tightrope. Words and Wisdom Carlsbad* (Carlsbad: Smiley Books, 2008), p. 5.
36 Dirk Hoyer, 'Mapping the Mindspace of Retopia: On Political Imagination', in *Strange Vistas: Perspectives on the Utopian, Mediated Fiction Series* (vol. 15), ed. by Ludmilla Gruzewska-Blaim. (Frankfurt am Main: Peter Lang, 2019).
37 Milton Friedman, 'The Role of Government in a Free Society', Free to Choose 1990 series, Hoover Institution (1990) <https://miltonfriedman.hoover.org/objects/57882/volume-5-created-equal>
38 Milton Friedman and Rose Friedman, *Free to Choose: A Personal Statement* (New York: Harcourt Brace Jovanovich, 1980), p. 27.
39 Anthony Giddens, *The Third Way: The Renewal of Social Democracy* (New York: Wiley, 1998), pp. 1–2.
40 Tony Blair and Gerhard Schröder, *Europe: The Third Way/Die Neue Mitte Friedrich Ebert Stiftung* (Johannesburg: Friedrich Ebert Foundation, 1998), p. 8.
41 Tony Blair, 'Labour Conference Speech 2005', *Guardian* (2005) <https://www.theguardian.com/uk/2005/sep/27/labourconference.speeches> [accessed 1 May 2022]
42 Giddens, pp. 35–36.
43 Simon Dein, 'Covid-19 and the Apocalypse: Religious and Secular Perspectives', *Journal of Religion and Health*, 60(1) (2021), pp. 5–15.
44 John J. Collins, *The Apocalyptic Imagination* (Grand Rapids: Eerdmans Publishing, 1984), p. 25.

45 Matthew Barret Gross and Mel Gilles, *The Last Myth – What the Rise of Apocalyptic Thinking is Telling Us about America* (New York: Prometheus Books, 2012), p. 239.
46 Michael Shermer, 'The End is Always Nigh in the Human Mind', *NewScientist*, 1 June (2011) <https://www.newscientist.com/article/mg21028156-300-the-end-is-always-nigh-in-the-human-mind/> [accessed 12 May 2022]
47 Matthias Riedl, 'Apocalyptic Politics: On the Permanence and Transformations of a Symbolic Complex', Humanities Futures Franklin Humanities Institute (2014) <https://humanitiesfutures.org/papers/apocalyptic-politics-on-the-permanence-and-transformations-of-a-symbolic-complex/> [accessed 12 May 2022]
48 Ibid.
49 Ibid.
50 Milton Friedman, *Capitalism and Freedom* (Chicago: University of Chicago Press, 1982, originally published 1962), p. viii.
51 William Faulkner, *Requiem for a Nun* (New York: Random House, 1951), p. 50.
52 Svetlana Boym, 'Nostalgia and Its Discontents', *The Hedgehog Review*, 9(2) (2007), p. 7.
53 Keith Naughton and Bill Vlasic, 'The Nostalgia Boom: Why the Old is New Again', *BusinessWeek*, March 23, 58–64 (1998), p. 60.
54 Boym, p. 15.
55 Ibid., p. 9.
56 Ibid., p. 8.
57 Fred Davis, *Yearning for Yesterday: A Sociology of Nostalgia* (New York: The Free Press, 1979), p. 33.
58 Ibid., pp. 34–35.
59 Zygmunt Bauman, *Retrotopia* (Cambridge: Polity, 2017).
60 Mark Lilla, *The Shipwrecked Mind: On Political Reaction* (New York: New York Review Books, 2016), p. xiv.
61 Boym, p. 12.
62 Nancy Fraser, 'The End of Progressive Neoliberalism', *Dissent Magazine* (2017) <https://www.dissentmagazine.org/online_articles/progressive-neoliberalism-reactionary-populism-nancy-fraser> [accessed 22 April 2022]
63 Fisher, p. 59.
64 Frederic Jameson, *The Seeds of Time* (New York: Columbia University Press, 1994), p. 1.
65 George Monbiot, *Bring on the Apocalypse: 6 Arguments for Global Justice* (London: Atlantic Books, 2009), p. 3.
66 Zygmunt Bauman, 'Symptoms in Search of an Object and a Name', in *The Great Regression*, ed. by H. Geiselberger (Cambridge: Polity Press, 2017), p. 21.
67 Ibid., p. 22.
68 Milton Friedman, *Neoliberalism and Its Prospects from the Collected Works of Milton Friedman*, compiled and edited by Robert Leeson and Charles G. Palm (Washington, DC: Farmand/Human Events, 1951), p. 2.
69 Immanuel Wallerstein, *Utopistics. Or, Historical Choices of the Twenty-First Century* (New York: The New Press, 1998), p. 4.

Bibliography

Arendt, Hannah, *The Human Condition* (2nd ed.). (Chicago: The University of Chicago Press, 1958)

Arendt, Hannah, *Lectures on Kant's Political Philosophy*. R. Beiner (ed.) (Brighton: Harvester, 1982)

Barret, Matthew and Gilles, Mel, 'How Apocalyptic Thinking Prevents Us from Taking Political Action', *The Atlantic*, April 23 (2012)

Barret Gross, Matthew and Gilles, Mel, *The Last Myth - What the Rise of Apocalyptic Thinking is Telling Us About America* (New York: Prometheus Books, 2012)

Bauman, Zygmunt, *Retrotopia.* (Cambridge: Polity, 2017)s

Bauman, Zygmunt, 'Symptoms in Search of an Object and a Name', in H. Geiselberger (ed.) *The Great Regression* (Cambridge: Polity Press, 2017)

Blair, Tony and Schröder, Gerhard, *Europe: The Third Way/Die Neue Mitte Friedrich Ebert Stiftung* (Johannesburg: Friedrich Ebert Foundation, 1998)

Blair, Tony, 'Labour Conference Speech 2005', *Guardian* (2005) <https://www.theguardian.com/uk/2005/sep/27/labourconference.speeches> [accessed 1 May 2022]

Bloch, Ernst, *The Principle of Hope* (reprint ed. 1995). Plaice, N., Plaice, S., and Knight, P. (trans.). (Cambridge, MA: MIT Press, Vol. 3:3, 1986)

Boym, Svetlana, 'Nostalgia and Its Discontents', *The Hedgehog Review* 9(2) (2007), pp. 7–18

Burke, Edmund, *Philosophical Enquiry into the Origin of Our Ideas of the Sublime and Beautiful* (London: R. and J. Dodsley, 1757)

Calhoun, Cheshire, *Motivating Hope. In Doing Valuable Time: The Present, the Future, and Meaningful Living* (Oxford: Oxford University Press, 2018)

Castoriadis, Cornelius, *Philosophy, Politics, Autonomy. Essays in Political Philosophy* (Oxford: Oxford University Press, 1991)

Castoriadis, Cornelius, *World in Fragments: Writings in Politics, Society and the Imagination* (Stanford: Stanford University Press, 1997)

Collins, John J., *The Apocalyptic Imagination* (Grand Rapids: Eerdmans Publishing, 1984)

Davis, Fred, *Yearning for Yesterday: A Sociology of Nostalgia* (New York: The Free Press, 1979)

Dein, Simon, 'Covid-19 and the Apocalypse: Religious and Secular Perspectives', *Journal of Religion and Health*, 60(1) (2021), pp. 5–15

Faulkner, William, *Requiem for a Nun* (New York: Random House, 1951)

Fisher, Mark, *Capitalist Realism: Is There No Alternative?* (Winchester, UK: O Books, 2009)

Fraser, Nancy, 'The End of Progressive Neoliberalism', *Dissent Magazine* (2017) <https://www.dissentmagazine.org/online_articles/progressive-neoliberalism-reactionary-populism-nancy-fraser> [accessed 22 April 2022]

Friedman, Milton, *Neoliberalism and its Prospects from The Collected Works of Milton Friedman*, compiled and edited by Robert Leeson and Charles G. Palm (Washington, DC: Farmand/Human Events, 1951)

Friedman, Milton, *Capitalism and Freedom* (Chicago: University of Chicago Press, 1982, originally published 1962)

Friedman, Milton, 'The Role of Government in a Free Society' in Free to Choose 1990 series (1990) <https://miltonfriedman.hoover.org/objects/57882/volume-5-created-equal>

Friedman, Milton and Friedman, Rose, *Free to Choose: A Personal Statement* (New York: Harcourt Brace Jovanovich, 1990)

Giddens, Anthony, *The Third Way: The Renewal of Social Democracy* (New York: Wiley, 1998)

Glassner, Barry, *The Culture of Fear: Why Americans Are Afraid of the Wrong Things: Crime, Drugs, Minorities, Teen Moms, Killer Kids, Mutant Microbes, Plane Crashes, Road Rage, & So Much More* (New York: Basic Books, 2001)

Goodwin, Barbara and Taylor, Keith, *The Politics of Utopia: A Study in Theory and Practice* (London: Hutchinson & Co, 1981)

Graeber, David, *Revolutions in Reverse. Essays on Politics, Violence, Art and Imagination* (New York: Minor Compositions, 2011)

Haiven, Max and Khasnabish, Alex, *The Radical Imagination: Social Movement Research in the Age of Austerity* (Halifax: Fernwood Publishing and London: Zed Books, 2014)

Herzler, Joyce O., *The History of Utopian Thought* (London: George Allen & Unwin, 2012, original work published 1923)

Hoyer, Dirk, 'Mapping the Mindspace of Retopia: On Political Imagination', in *Strange Vistas: Perspectives on the Utopian*, ed. by Gruzewska-Blaim, Ludmilla, Mediated Fiction Series (vol.15). (Frankfurt am Main: Peter Lang, 2019)

Jameson, Frederic, *The Seeds of Time* (New York: Columbia University Press, 1994)

Levitas, Ruth, *The Concept of Utopia* (Witney: Peter Lang, 2011)

Levitas, Ruth, *The Imaginary Reconstitution of Society: Or Why Sociologists and Others Should Take Utopia More Seriously*, Inaugural Lecture. (University of Bristol, 2005) <http://www.bristol.ac.uk/depts/sociology/staff/pubs/levitasinaugural> [accessed 12 February 2019]

Lilla, Mark, *The Shipwrecked Mind: On Political Reaction* (New York: New York Review Books, 2016)

McKinnon, Catriona, 'Cosmopolitan Hope', in *The Political Philosophy of Cosmopolitanism*, ed. by G. Brock and H. Brighouse (Cambridge: Cambridge University Press, 2005), pp. 243–249.

Monbiot, George, *Bring on the Apocalypse: 6 Arguments for Global Justice* (London: Atlantic Books, 2009)

More, Thomas, *Utopia* (New York: Dover Publications, 1997)

Mouffe, Chantal, *On the Political* (London: Routledge, 2005)

Mulgan, Geoff, 'The Imaginary Crisis (and How We Might Quicken Social and Public Imagination)', *UCL Demos Helsinki and Untitled* (2020) <https://demoshelsinki.fi/julkaisut/the-imaginary-crisis-and-how-we-might-quicken-social-and-public-imagination/> [accessed 23 April 2022]

Naughton, Keith and Vlasic, Bill, 'The Nostalgia Boom: Why the Old is New Again', *BusinessWeek*, March 23, 58–64 (1998)

Pettit, Philip, 'Hope and It's Place in the Mind', *The Annals of the American Academy of Political and Social Science*, 592 (2004)

Riedl, Matthias, 'Apocalyptic Politics: On the Permanence and Transformations of a Symbolic Complex', Humanities Futures Franklin Humanities Institute (2014) <https://humanitiesfutures.org/papers/apocalyptic-politics-on-the-permanence-and-transformations-of-a-symbolic-complex/> [accessed 12 May 2022]

Rioux, Catherine, 'Hope: Conceptual and Normative Issues', *Philosophy Compass*, 16(3) (2021)

Seargeant, Philip, *The Art of Political Storytelling* (London: Bloomsbury Academic, 2020)

Shermer, Michael, 'The End is Always Nigh in the Human Mind', *NewScientest*, 1 June (2011) <https://www.newscientist.com/article/mg21028156-300-the-end-is-always-nigh-in-the-human-mind/> [accessed 12 May 2022]

Taylor, Charles, *Modern Social Imaginaries* (Durham, NC: Duke University Press, 2004)

Taylor, Charles, *Secular Age* (Cambridge, MA: Harvard University Press, 2007)

Wallerstein, Immanuel, *Utopistics. Or, Historical Choices of the Twenty-First Century* (New York: The New Press, 1998)

Watts, Jonathan, 'Interview: Greta Thunberg, Schoolgirl Climate Change Warrior: 'Some People Can Let Things Go. I Can't'', *The Guardian*, March (2019) <https://www.theguardian.com/world/2019/mar/11/greta-thunberg-schoolgirl-climate-change-warrior-some-people-can-let-things-go-i-cant> [accessed 20 January 2021]

West, Cornel, *Hope on a Tightrope. Words and Wisdom Carlsbad* (Carlsbad: Smiley Books, 2008)

2 Retelling the Story

Five Premises

Consider the basic premises of these five stories:

1) a protagonist founds a start-up business and works her/his way up from poverty to success
2) a protagonist refuses to work and to interact with the world in its totality
3) an organized group of people from all over the planet plans and executes the world revolution
4) a disorganized group of people from all over the planet roams through the forest in search of a rare and valuable mushroom in order to secure their survival.
5) human beings dissolve, enter as equals into a multispecies universe, and become critters among critters in a permanent process of making and unmaking in the Chthulucene

If you could choose to be the protagonist in any of the five stories, which one would you choose? Marshall Ganz developed for Barack Obama's 2008 campaign the Public Narrative Method in which participants have to define what, for them, is the story of self, the story of us, the story of now. So, which of these stories is the one that, for you, best reflects your ideas of self, us, and now?

A closer look at these premises reveals that they tell stories that have a cultural impact on the present time. Premise 1 sums up the Milton Friedman type story of neoliberalism. Premise 2 is the Bartleby type story, so the Herman Melville story reinterpreted by philosophers such as Giorgio Agamben, Gilles Deleuze, or Byung-Chul Han ('I would prefer not to'). Premise 3 reflects the story of Marxist world revolution. Premise 4 tells Anna Tsing's story of the global Matsutake harvest and trade. And, lastly, Premise 5 tells us the story of Donna Haraway's Chthulucene.

As Philip Seargeant underlines: 'Storytelling is an essential element in the way we interpret the entire social world. [...] The tales we tell not only

DOI: 10.4324/9780429288180-2

shape the times we live in but also reflect them'.[1] How do these four stories thus shape our present era and how do they reflect them?

In order to understand the impact of stories, Patrick Reinsborough and Doyle Canning have developed the narrative power analysis. The method is influenced by the Gramscian notion of cultural hegemony. Narrative power analysis is thus based on the recognition that 'that humans understand the world and our role in it through stories, and thus all power relations have a narrative dimension'.[2] These stories evidently also either challenge or justify the status quo. Reinsborough and Canning propose five criteria of story analysis: the conflict (what does the story present as a problem), the character (who is (s)he? How relatable is (s)he?), the imagery (what is shown), foreshadowing (how does the story hint at its outcome), and the assumptions.

The assumptions are central for the reception of the story. According to Reinsborough and Canning, the assumptions 'make up the glue that holds the story together; they are the unstated parts of the story that you have to accept in order to believe the narrative is true'.[3]

In order to analyze the narrative dimensions of power, the concept of story needs to be defined. For the sake of this analysis two definitions of story offer useful insights. Robert McKee defines classical story design as:

> a story built around an active protagonist who struggles against primarily external forces of antagonism to pursue his or her desire, through continuous time, within a consistent and causally connected fictional reality to a closed ending of absolute irreversible change.[4]

McKee calls these timeless principles 'archplot'. Lisa Cron's definition is also referring to the traditional definition of linear narratives when she claims that a story 'is about how the things that happen affect someone in pursuit of a difficult goal, and how that person changes internally as a result'.[5] While these definitions obviously do not incorporate the totality of story variations, they reflect the organization principles of most human experiences. As Paul Ricoeur explains 'time becomes human time to the extent that it is organized after the manner of a narrative; narrative, in turn, is meaningful to the extent that it portrays the features of temporal experience'.[6]

Robert McKee also emphasizes the universal appeal of continuous narratives. McKee points out that:

> Most human beings believe that life brings closed experiences of absolute, irreversible change; that the greatest sources of conflict are external to themselves, that they are the single active protagonists of their own existence; that their existence operates through continuous time within

> a consistent causally interconnected reality; and that inside this reality events happen for explainable and meaningful reasons.[7]

While this claim can be contested in principle, it offers a good foundation for narrative analysis.

What are thus the narrative dimensions of power in the four premises: the Friedmanian story (Premise 1), the Bartleby story (Premise 2), the Marxist story (Premise 3), the Anna Tsing Story (Premise 4), and the Chtlhulucene story (Premise 5)?

The Friedmanian Story

The Friedmanian story presents the conflict as an external conflict (poverty) which can be overcome by hard work, entrepreneurial spirit, and the right choices. The archetypical version of this story precedes Milton Friedman: the American dream. Poverty is thus defined in materialistic terms, and the conflict can be overcome by the accumulation of financial wealth. But the fundaments of this story are even older, as it reflects what Christopher Booker defines as one of the seven basic plots. The Friedmanian story is thus an update of the conflict of the American Dream, which is a reflection of the ancient plot 'from rags to riches'. The protagonist has to fight against all the forces that stop her on her way to upward mobility. The antagonists can potentially be multifold: unsupportive parents, envious colleagues or friends, narrow-minded bosses, skeptical investors... The premise taps into a basic storytelling pattern, and the conflict reflects the transhistorical desire to accumulate wealth. But tapping into the transhistorical form of the conflict camouflages the fact that, in the beginning of the 21st century in Western society, upward mobility is drastically on the decline. Especially after three decades of neoliberalism, the prevailing itinerary is 'from riches to riches'. But the 'from riches to riches' plot would lack any conflict and thereby make a stale and boring story.

The leading character in the Friedmanian story is an individual with the strong aspirations for wealth. Material poverty drives her into her hero's journey. She is per definition active, with a clear goal (get rich), and the forces of antagonism are external. In order to reach her goal, she has to overcome obstacles that challenge her to the maximum. In the end she prevails because she has overcome many crises and deserves her heroine status and is thus rewarded with financial wealth. The money is the reward for her journey. But it is not the only reward: her inner change made her a stronger person. This is validating the idea that wealth is something that needs to be earned through hard struggle. It is a sign of a strong character, the reward for the hero's journey, and therefore deserves to be respected.

The imagery of such stories often associates material poverty with mental poverty and material wealth with new horizons and new options. So, the dirty and dark studio in the Bronx is contrasted with the top floor views and the lights of Midtown Manhattan, the council houses of Tower Hamlets are contrasted with the Georgian townhouses in Mayfair, and the garbage-littered puddles of Kusumpur Pahariwith with the pools of Prithviraj Road. The implicit assumption is that the poor inhabitants in underprivileged neighbourhoods are stuck there because they lack the exceptional qualities of the protagonist. The visual point of departure, the derelict neighbourhood, is thus to be left behind in favour of the beauty and the promise of happiness in the wealthy parts of town.

The Friedmanian story foreshadows the idea that exceptional effort will be also exceptionally rewarded. When the protagonist accepts the struggle, the drawbacks, and the pain and still continues on her path, it hints at the idea that the reward will be justified. The conflict can thus be resolved by commitment to the cause of individual upward mobility. The heroine has helpers but ultimately accomplishes her goal through her own strength.

The underlying assumption of the 'rags to riches' story in its Friedmanian version is that effort and hard work will be rewarded with wealth. This fundamental belief is the control myth of neoliberalism. It taps into the basic pattern of human hope for improvement and reduces it to the idea of material wealth.

The Bartleby Story

The Bartleby premise is based on Herman Melville's short story 'Bartleby: A Story of Wall Street'. The narrator of the story is a Wall Street lawyer who describes his employee Bartleby. The scrivener, who is initially very hard-working, gradually declines any request by his boss with the sentence 'I would prefer not to'. Bartleby refuses to work, refuses to leave the office, and refuses any kind of change. When the lawyer moves his business to another building, he leaves Bartleby behind since he prefers not to change anything. The new tenant of the building evicts Bartleby, and the police arrest him as a vagrant. The narrator visits Bartleby in prison and attempts to improve his conditions there but, on his last visit, he finds out that Bartleby died of starvation. He rejected any food with a polite 'I would prefer not to'. Bartleby's obstinate refusal of action inspired Derrida, Deleuze, and Agamben, who stated that the scrivener's action 'calls into question this supremacy of the will over potentiality'.[8] Byung-Chul Han interprets Bartleby as a degraded *animal laborans* who 'is not burdened by the late-modern imperative of letting his *self* flourish'.[9]

Bartleby's conflict is largely an inner conflict, and since he does not say anything other than his trademark sentence, a lot of room is left for speculation (which is probably one of the reasons why this story attracts so much interpretation from philosophers). It is clear that the working conditions and the monotony of his tasks are contributing to Bartleby's refusal to engage with the world. The Wall Street lawyer narrates the story and presents himself as a supportive and humane boss. Thus he removes his possible role as antagonist to Bartleby from the narrative. The outside world is presented as antagonistic, chiefly the new landlord and the legal system. The interesting feature of the conflict in the Bartleby story is the reaction of the protagonist to the conflict. Bartleby prefers not to act.

The story is thus based on a passive protagonist who does not express (verbally or nonverbally) a manifest goal. The narrator thus has to serve as a messenger of the story and, while he presents contextual information, it is impossible to make a definite statement of the character's goal, his motivation, his wants, his needs, his desire. Bartleby is thus not just a passive protagonist but also an obscure protagonist. There is no hero journey because the protagonist refuses to take any role in the narrative. Bartleby's silent immobility is challenging traditional modes of identification with his character. It does not permit the audience to assess the inner change of the protagonist. Bartleby remains a stranger.

As Byung Chul-Han pointed out, the imagery of the story is bleak, the lawyer's office is the grey environment for the degraded animal laborans. Once Bartleby is arrested by the police and ends up in prison, his existence becomes even more degraded. In a repressive prison system the only possibility to opt out is to starve to death.

Due to the manifestly passive Bartleby, the story does not foreshadow the future, it does not show any vision to resolve the conflict since the protagonist's polite 'I would prefer not to' does not offer any potential for development.

The underlying assumption is thus that it is possible to reject the (labor) world, that obstinate resistance can be a valid attitude to undermine the functioning of the system. But, unlike in the interpretations by Agamben and Deleuze, in Melville's story the outcome of such an attitude is prison and death. 'I would prefer not to' does not undermine the mechanisms of a dehumanizing system but leads to a sacrifice without consequences. The scrivener's story, in Christopher Booker's 'seven basic plots system',[10] is thus a tragedy.

The Revolution Story

In the Marxist premise, the world revolution has been staged in world history on different scales. The biggest revolutions, in France (pre-Marx), Russia, and China, had a profound impact on history, even if they never evolved into world revolutions.

Marx perceived capitalist society in antagonistic ways, as a conflict between capitalists and the working class. So, the conflict structure of the Marxist story is clear and simple: on the one hand, the greedy antagonist, the capitalist; on the other, the exploited protagonists, the working class.

As the history of pre-Stalin Soviet cinema has shown, communist storytelling, in the beginning, was based on the masses as protagonists. The individual hero, despite the glorification of leaders like Lenin, was seen as a bourgeois construct of Hollywood films. So, the lead character in the Marxist story is a whole social class. Through their unity and desire for change, these multiple protagonists actively pursue their goal (which is difficult to achieve): to beat the villain by any means possible. As a result, the protagonists not only change internally (achieve freedom) but also change the external framework (the society).

The foreshadowing is based on the idea that a continuous struggle led by a mentor (the communist party) is leading to a successful reversal of the economic system, and the capitalist system will be replaced by a society that is based on the wants and needs of the underprivileged. This foreshadowing, combined with the leap of faith that revolutionary change on a global scale is achievable, provided the motivation to continuously engage with the struggle.

The underlying assumption of the Marxist premise is that everything is justified in order to realize the goal (the end justifies the means). The bloodshed of Stalin or Mao era communism are only the most visible aberrations. The other underlying assumption is that there is a unified class with a common consciousness that is capable of maintaining group coherence over a long time and challenging the capitalist system. The post-World War II developments in Western societies have shown that fragmentation has dissolved Marx's protagonists of revolution: the working class has left the world stage. The third underlying assumption is that the villain of the story, capitalism, is an immutable unit that can be easily identified and then combatted. As it turned out, the 'villain' is adaptable and not only financially potent but also apt in psychological manipulation.

The basic plot of the Marxist story, 'Overcoming the Monster', thus led to a story in which 'the monster' overcame the heroes. Despite the failure of the story, the trope of revolutionary change is still attractive for many freedom movements. The underlying desire for a radical removal of a dysfunctional system is still strong. But, as history has shown, revolutions happen in societies with expansive demographic pyramids, for instance, countries with many young people. Europe has a constrictive demographic pyramid. Aging continents will not be the stage for future revolutions. So, the Marxist premise, at least in the European context, might be rather an expression of the desire to be young than the premise of a story that will lead to the reversal of the status quo.

The Matsutake Story

The Matsutake premise is based on Anna Tsing's description of a global network of people that are engaged with the harvest and trade of the most expensive mushroom in the world. The matsutake mushroom is considered to be the most distinguished gift in Japan, but due to the changes in the environment, it does not grow in the country anymore. In order to satisfy the need of the traditionally minded Japanese clientele, in diverse places around the world, people forage the mushrooms and sell them to Japan. The interesting feature of the matsutake mushroom is that it grows in ecologically damaged places. It is hard to find and thus difficult to harvest which makes the product not scalable in economic terms and does not permit capitalist exploitation in the classic sense. Tsing, for instance. describes how groups of Lao refugees in Oregon have created their livelihood based on picking matsutake mushrooms. The matsutake chain of supply, in Tsing's interpretation, is thus a precarious arrangement of collaborations which are organized on the edge of capitalist value creation, which includes a form of multispecies coexistence.

The conflict of the Matsutake premise is the situation of economic precarity and environmental destruction that creates the need for new livelihoods for an increasing part of the world population. As Tsing explains:

> Precarity is the condition of being vulnerable to others. Unpredictable encounters transform us; we are not in control, even of ourselves. Unable to rely on a stable structure of community, we are thrown into shifting assemblages, which remake us as well as our others. We can't rely on the status quo, everything is in flux, including our ability to survive.[11]

The universe of the Matsutake premise is thus shifting, in flux, in permanent reconfiguration. While the antagonist is still the same as in the Marxist premise (capitalism) there is no way to fight the antagonist. The only thing left to the protagonists is thus to develop a strategy of survival. Unlike the Marxist story, there is also no stable group to fight the antagonist anymore, so the assemblages of people and situations are permanently shifting.

The protagonist is thus in a precarious situation in a world of flux. The cause-and-effect scheme is unstable, and the external forces of antagonism are difficult to locate. The protagonist in the Matsutake story has to be active without being able to articulate a goal other than (multispecies) survival. But unlike an adventure story, the outer forces are difficult to identify, and it is impossible to tell when the battle is won. In addition, the protagonist cannot rely on his individual mind, and not even on his anthropocentric mind, to develop strategies of survival. The internal change in the ongoing story is thus the acknowledgement of ontological precarity. In the absence of a clearly identifiable external antagonist, the antagonist is the self (at least in the limited individualist setup). Thus, there are multiple protagonists in the Matsutake premise, including nonhuman actors.

The imagery of the Matsutake story is global: forests in Oregon, Yunnan, Lapland, and trading places in Japan. People of different cultural backgrounds with different motivations, different traditions, and different interests are all linked by a fungus. The juxtaposition of wilderness and civilization, combined with the cultural diversity of the participants, creates an imagery that alludes to diversity.

Each of the stories foreshadows the idea of precarity. The conclusion is foreclosed, the ending is open, everything is precarious. If causality is replaced by coincidence, it is impossible to foreshadow future developments. Consequently, the realities in the Matsutake story are inconsistent and nonlinear.

As for the underlying assumptions of the story: the Matsutake premise is an acknowledgement that it is impossible to tell a linear story. As Anna Tsing declared:

> I find myself without the handrails of stories that tell where everyone is going and, also, why. Precarity once seemed the fate of the less fortunate. Now it seems that all our lives are precarious even when, for the moment, our pockets are lined. [...] I address the imaginative challenge of living without those handrails, which once made us think we knew, collectively, where we were going.[12]

The underlying assumption is thus that it is impossible to tell a collective story in times of precariousness. Tsing abandons the handrails of stories and instead postulates the imaginative challenge to live without them. As an archetypal plot, the Matsutake story, can be described as a quest without narrative closure.

Tsing's analysis is based on the postulate that only in the 21st century is life precarious for the majority of the population. This limited perception is blind to all the periods of precariousness humans had to experience in history. From a point of view of narrative power analysis, the Matsutake story (despite being fascinating) is weak because it renounces the potential power of stories to stimulate the collective imagination. In the absence of collective imagination ('the story of us'), the danger is that imagination reverts into the very state that Tsing attempts to overcome: the individual imagination.

The Chthulucene Story

The Chthulucene premise by Donna Haraway is also based on the intention to overcome the individual imagination. Haraway's concept of sympoiesis is based on tentacular thinking, interspecies entanglement, and multispecies flourishment based on 'making with'. She points out that 'bounded individualism in its many flavors in science, politics and philosophy has finally become unavailable to think with, truly no longer thinkable, technically or any other way'.[13] As an alternative, Haraway proposes that 'human beings dissolve, enter as equals into a multispecies universe, become critters among critters in a permanent process of making and unmaking in the Chthulucene'.[14]

The Chthulucene (derived from the Greek 'khtōn' and 'kainos') is based on the idea that the chthonic ones (the beings under the earth) are in a permanent process of making and unmaking in the 'kainos' (the now, unbound by past or future). This process is undoing the Capitalocene and the Anthropocene that have contributed to the quasi total destruction of the livelihood on the planet. Multispecies kinship is based on tentacular thinking which is not visual and anthropomorphic but reachable through the senses of touch and feel and open experimentation. Humans should thus learn from spiders like Pimoa Cthulhu, with their tentacular approach to the earth. Haraway acknowledges the human potential to be humus and thus states: 'Human as humus has potential, if we could chop and shred human as Homo, the detumescing project of a self-making and planet-destroying CEO'.[15]

This reverts to the classic scheme of disaster films. If aliens, monsters, dinosaurs, or other creatures usually are presented as the source of conflict for the human protagonists, then the Chthulucene premise presents the human protagonists as a menace to the flourishing of the multispecies world. Giant spiders are not a menace but an inspiration.

The human species is cast as a villain in this premise, and the critters are the heroes. The villain has to change into a critter in order to engage in a sympoietic entanglement with the other critters. The internal change of the character thus entails in refraining from human rationality, human thinking, and thus human perception.

The imagery is a compost pile full of endless composing and recomposing creatures that are engaged in permanent entanglement with the muddy earth. Everything moves, everthing is chaos, everything is, nothing has direction or ontological meaning. The Chthulucene compost pile has no past, no present, no future, it just recomposes and re-composts itself on a permanent basis. The *homo sapiens* is thus thrown into a pile of biological recomposition in a sort of reverse evolution.

The question of foreshadowing does not make any sense in this context because, in the Chthulucene, the Pimoa Cthulhu and the (former) homo sapiens is a probing creature without any sense of future.

Under the lens of Booker's archetypal plot lines, it is unclear whether this scenario is a comedy, a voyage and return, a rebirth, or a tragedy.

The underlying assumption of the Chthulucene, the possibility of a metamorphosis of the homo sapiens into a critter, is thought-provoking. Haraway calls to abandon the anthropomorphic and visual thinking. As a playful critique of the Capitalocene and the Anthropocene, the Chthulucene presents an alternative. But other than that, there are too many flaws in logical coherence (logic can be a chthonic critical kill-joy): why is Donna Haraway writing articles and books as opposed to joyfully interacting as a critter in a compost pile? The answer: we cannot think away our own human embodied existence. Critters are beautiful creatures but even if the anthropo-centred worldview has caused so much damage they are not providing the answer to our human existence. To think away the human protagonists from the human story means to foreclose the possibility that the human imagination (the one that is not blinded by what Haraway calls the 'comic faith in techfixes'[16]) can find other possibilities for multispecies coexistence than becoming a critter. From a narrative power analysis, the Chthulucene premise is thus not providing a counter-narrative with the potential to challenge the control mythology. If the critical part of humanity decides to mimic the Pimoa Cthulhu and cheerfully engage with a compost pile, then the CEOs of the Capitalocene will have even less resistance in their exploitation of the human and natural resources.

New Premises for the Story Wars

If the neoliberal ego-shooters, the shoe-gazing Bartlebys, the failed revolutionaries, the Matsutake foragers, and the chthonic critters cannot provide a premise for a story that can challenge the control mythology of our time, then how can we find a premise for a new story?

Jonah Sachs describes stories as a type of human communication that is designed to persuade the audience of the storyteller's worldview. In turn, the storyteller's worldview resonates (or not) with the audience's worldview. Sachs defines worldview as 'a collection of stories about how things have come to be the way they are and what we should do about them'.[17] But increasingly, the realities of the lived moment are not in tune anymore with the shared stories. Sachs thus diagnoses a 'myth gap' and points out:

> We live in a world that has lost its connection to its traditional myths, and we are now trying to find new ones – we're people and that's what people without myths do. These myths will shape our future, how we live, what we do, and what we buy. They will touch all of us. But not all of us get to write them. Those who do have tremendous power.[18]

The myth gap is often filled with the marketing aspects of global brands who, in Sachs' description, use two basic strategies to persuade consumers. Inadequacy marketing is based on fear and doubt (for example, 'you are not beautiful enough, use our face cream, and then you will be attractive'). Empowerment marketing is based on the positive appeal of a product to enhance the possibilities of the consumer (for example, 'buy these sport shoes, go jogging, excel, become active and healthy, just do it'). This can be also transferred to political marketing and storytelling. But these stories are micronarratives that are quickly replaced by new micronarratives. Brands thus fill the myth gap but with the specific purpose to sell their product, be it a sports shoe or a presidential candidate.

In what Jonah Sachs calls 'the story wars', the essence is that each story is in competition with the others. Hence the narrative power is a decisive element. Stories that shape a worldview are thus at the core of society's creation of norms and values, or as Sachs explains it: 'Shared stories, shared values, shared worldview – us'. In the battle for defining the big stories in a society, the old dictum that without conflict there is no story can be also reverted: without a story, there is no conflict.

Sachs proposes to use the methods of empowerment marketing with the aim of 'reorienting our myth away from the adolescent, depressed consumer mind-set and toward the empowered citizen worldview'.[19] But, of course, the concept of empowered citizenship can be co-opted for all political purposes. Two political movements that successfully tapped into the ideas of empowerment marketing were the Tea Party and Occupy Wall Street. As Sachs points out:

> Both of these movements worked because they stepped into the gap left behind by the fraying myth of the American Dream to supply a new story (they provide clear character and conflict), explanation (they tell us the root cause of the meltdown is simple), meaning (exhort us as average people to stand up) and even ritual (they invite us to tea parties and tent cities).[20]

Both movements were reflecting the unsurmountable myth gap caused by the 2008 economic crisis, provided a new story (including tapping into powerful myths of the past, 'the Boston Tea party', or the present, 'occupying spaces as a revolutionary act'), explanation, and meaning. But, more than a decade later, both movements have disintegrated. The same has happened to the Indignado movement in Spain, to the Indignant Citizens movement in Greece, to protest movements in Hong Kong, Chile, Iraq, Lebanon, Ukraine, Egypt…

Reinsborough and Canning underline that mobilization when the myth gap is painfully obvious to large parts of the population is different from the idea of creating a sustainable, alternative new story. They distinguish the story of the battle from the battle of the story and point out: 'The story of the battle is about mobilization; the battle of the story is about persuasion'.[21]

Protest movements are largely operating with the story of the battle, which is based on shared assumptions (e.g. about the abusive power of corporations, the corruption of the political system or specific demands of human rights), a common tactical approach (e.g. the occupation of a square), and a partisan story of what is happening and what needs to happen. Through the effective use of social media, the story of the battle can result in the mobilization of millions of protesters, for example the Women's March in 2017, the Black Lives Matter protests in 2020, or the Indian farmer protests in 2021. In many cases, though, the euphoria of the success of the mobilization campaigns is not long-lasting because the fundamental political arrangements stay the same despite the protests.

The more fundamental challenge is thus the battle of the story which is the larger struggle to determine whose stories are told, what stories are told and how stories are framed (all of which is essentially what Jonah Sachs describes as 'story wars'). Essentially, the battle of the story is the long-term strategy to gain narrative power and occupy the vacant space in the myth gap (as opposed to the tactical and momentary establishment of a protest micro-narrative).

Once the difference between the tactical story of the battle and the strategical battle of the story is sorted out, the real challenge is to tell a story that has the power to occupy the vacant space in the myth gap. David Ricci calls this story an alpha story which he defines as large and lasting tales that 'are powerful enough to trump competing tales [...] within a political camp and which can then compete against alpha stories from the rival political camps'.[22] He sees a reluctance to 'alpha stories' in the left and liberal political camp, an audience problem. Ricci observes: 'Liberals prefer not to adopt stories and act accordingly even when, spot on, someone spins for them a tale forcefully and eloquently'.[23] In addition to the audience problem, Ricci is pessimistic about the possibility on either side of the political spectrum to create new alpha stories. As a consequence, conservatives keep telling the same old stories and liberals refuse to engage in what Reinsborough and Canning call the 'battle of the story' in what Jonah Sachs describes as 'story wars'.

Even if the market-based approach to storytelling has been criticized, for instance by Sujatha Fernandes, who deplores that 'storytelling in a neoliberal era comes to be guided by utilitarian logics',[24] to renounce from engaging with the battle of the story would mean foreclosing the possibility to redefine the rules of the game with a new story. So, on one level there is an audience problem with the highly self-reflexive liberals and leftists; on another level, there is an audience problem with..., well, the audience in total.

Alessandro Portelli, in his studies of oral history, comes to the conclusion that the subjectivity of the speaker rules supreme, what audiences (which he calls informants) 'believe is indeed a historical fact (that is, the fact that they believe in it) as much as what really happened'.[25] What he relates to as the difference of oral history to written history could also be described as the basic perceptive pattern towards events. Portelli claims that events matter less than their meaning, and thereby the interpretation of these events in the minds of the audience.

Somewhat ironically, any story in the story wars faces one important antagonist: the resistance of the human brain to change. Will Storr claims that the reason for this resistance to change is our heroic self-perception. Storr observes:

> if we're psychologically healthy, our brain makes us feel as if we're the moral heroes at the centre of the unfolding plot of our lives. Any 'facts' it comes across tend to be subordinate to that story. If these 'facts' flatter our heroic sense of ourselves, we're likely to credulously accept them, no matter how smart we think we are. If they don't our mind will find some crafty way of rejecting them.[26]

Based on neuro-scientific observations, Storr describes the process of narrative self-making as a process in which identities are formed around values and a personal theory of control (how to achieve what one wants). This process in most cases is completed after adolescence, and then brain switches its mode of perception. Storr describes this moment of change as a transformation from 'model-builders' to 'model defenders'.[27] The model we are defending is nothing more than the creation of our own minds, since, according to Storr, 'we're all fictional characters'.[28] The brain thus fights ferociously to defend our self-created fictional character, its norms, values, and identity. Despite the fact that Storr acknowledges that we are constantly altered by other people and that our brain is fascinated by change, changing ourselves fundamentally is a monumental challenge. The flaws that have entered the model-building process in the first two formative decades define and distort our perception of reality. As Storr emphasizes:

> Identifying and accepting our flaws, and then changing who we are, means breaking down the very structure of our reality before rebuilding it in a new improved form. We often fight with all we have to resist this kind of profound change. This is why we call those who manage it 'heroes'.[29]

Storr's descriptions of human beings as model defenders of their own heroic story which makes deeper change – and thus, in this definition, 'real heroism' – difficult to achieve, also casts a shadow on the potential for collective change.

The stories of the battles show that the desire for change is very present, but it often evaporates after a euphoric moment of protest. Without a story that shows a clear conflict, an active character, and an attractive imagery, without a story that foreshadows real change, without a story that has the underlying assumption that human agency is possible not only in the abstract, the model-defending hero brain will convert back into the mode of individualist profit maximization.

The five premises whether they are the Friedman premise, the Bartleby premise, the Marxist premise, the Matsutake premise, or the Chthulucene premise do not have the narrative power to challenge the model-defending hero brain. One important task is thus to develop new premises for stories that have the potential to do so. Refraining from engaging in the battle of the story with new premises would mean to leave the mind scape to either political inadequacy marketing (and spiral the fear to new levels) or to the reductive consumerist version of empowerment marketing.

One option might be to develop a premise that is based on a local scale, has a practical and active approach to community-building, leaves the freedom to define the political priorities within the community, and thus projects empowerment on a practical 'lived experience' level. Let's call this the 'retopian' approach.

Consider the retopian premise:

- a group of people from all over the world settles in a (half-) abandoned village in the periphery and designs a world according to their values.

The story of this book is based on this premise.

Notes

1 Philip Seargant, *The Art of Political Storytelling* (London: Bloomsbury Academic, 2020) pp. 6–7.
2 Patrick Reinsborough and Doyle Canning, *Re:Imagining Change* (Winnipeg: Fernwood Publishing, 2010), p. 20.
3 Ibid., p. 39.
4 Robert McKee, *Story: Substance, Structure, Style and the Principles of Screenwriting* (Sydney: Random House, 1997), p. 45.
5 Lisa Cronx, *Story Genius: How to Use Brain Science to Go Beyond Outlining and Write a Riveting Novel* (New York: Random House, 2016), p. 17.
6 Paul Ricoeur, *Time and Narrative*, vol. 1 (Chicago: University of Chicago Press, 1983), p. 3.
7 McKee, p. 62.
8 Giorgio Agamben, 'Bartleby, or On Contingency', in *Potentialities: Collected Essays in Philosophy*, trans. and ed. D. Heller-Roazen (Stanford: Stanford University Press, 1993) <https://web.english.upenn.edu/~cavitch/pdf-library/Agamben_Bartleby.pdf>, pp. 243–303 (p. 270) [accessed 5 October 2019]
9 Byung-Chul Han, *The Burnout Society* (Redwood City: Stanford University Press, 2015), p. 26.
10 Christopher Booker, *The Seven Basic Plots: Why We Tell Stories* (New York/London: Bloomsbury Continuum, 2004).
11 Anna Tsing, *The Mushroom at the End of the World* (Princeton: Princeton University Press, 2015), p. 20.
12 Ibid., p. 2.
13 Donna J. Haraway, *Staying with the Trouble - Making Kin in the Chthulucene* (Durham: Duke University Press, 2016), p. 5.
14 Ibid., p. 5.
15 Donna J. Haraway, 'Tentacular Thinking: Anthropocene, Capitalocene, Chthulucene', *e flux Journal*, Issue # 75, September (2016) <https://www.e-flux.com/journal/75/67125/tentacular-thinking-anthropocene-capitalocene-chthulucene/> [accessed 2 February 2020]
16 Donna J. Haraway, *Staying with the Trouble - Making Kin in the Chthulucene* (Durham: Duke University Press, 2016), p. 3.
17 Jonah Sachs, *Winning the Story Wars* (Boston: Harvard Business Press, 2012), p. 20.
18 Ibid., p. 6.
19 Ibid., p. 119.
20 Ibid., p. 78.
21 Reinsborough and Canning, p. 43.
22 David Ricci, *Politics Without Stories - The Liberal Predicament* (New York: Cambridge University Press, 2016), p. 24.
23 Ibid., p. 191.
24 Sujatha Fernandes, *Curated Stories: The Uses and Misuses of Storytelling* (New York: Oxford University Press, 2017), p. 10.
25 Alessandro Portelli, 'What Makes Oral History Different', in *The Oral History Reader*, ed. by R. Perks and A. Thomson (New York: Routledge, 2006), p. 36.

26 Will Storr, *The Science of Storytelling: Why Stories Make Us Human and How to Tell Them Better* (London: Harper Collins, 2019), p. 3.
27 Ibid., p. 85.
28 Ibid., p. 92.
29 Ibid., p. 65.

Bibliography

Agamben, Giorgio, 'Bartleby, or On Contingency', in trans. and ed. by D. Heller-Roazen, *Potentialities: Collected Essays in Philosophy* (Stanford: Stanford University Press, 1993) <https://web.english.upenn.edu/~cavitch/pdf-library/Agamben_Bartleby.pdf> [accessed 5 October 2019]

Booker, Christopher, *The Seven Basic Plots: Why We Tell Stories* (New York/London: Bloomsbury Continuum, 2004)

Brooks, Jerome, 'Chinua Achebe, The Art of Fiction' # 139, *The Paris Review*, Issue 133 Winter (1994) <https://www.theparisreview.org/interviews/1720/the-art-of-fiction-no-139-chinua-achebe> [accessed 23 May 2020]

Cron, Lisa, *Story Genius: How to Use Brain Science to Go Beyond Outlining and Write a Riveting Novel (Before You Waste Three Years Writing 327 Pages that Go Nowhere)* (New York: Random House, 2016)

Fernandes, Sujatha, *Curated Stories: The Uses and Misuses of Storytelling* (New York: Oxford University Press, 2017)

Han, Byung-Chul, *The Burnout Society* (Redwood City: Stanford University Press, 2015)

Haraway, Donna J., *Staying with the Trouble – Making Kin in the Chthulucene* (Durham: Duke University Press, 2016)

Haraway, Donna J., 'Tentacular Thinking: Anthropocene, Capitalocene, Chthulucene', *e flux Journal*, Issue # 75, September (2016) <https://www.e-flux.com/journal/75/67125/tentacular-thinking-anthropocene-capitalocene-chthulucene/> [accessed 2 Februray 2020]

McKee, Robert, *Story: Substance, Structure, Style and the Principles of Screenwriting* (Sydney: Random House, 1997)

Portelli, Alessandro, 'What Makes Oral History Different', in ed. by R. Perks and A. Thomson, *The Oral History Reader* (New York: Routledge, 2006)

Reinsborough, Patrick and Canning, Doyle, *Re:Imagining Change* (Winnipeg: Fernwood Publishing, 2010)

Ricci, David, *Politics Without Stories – The Liberal Predicament* (New York: Cambridge University Press, 2016)

Ricoeur, Paul, *Time and Narrative*, vol. 1 (Chicago: University of Chicago Press, 1983)

Sachs, Jonah, *Winning the Story Wars* (Boston: Harvard Business Press, 2012)

Seargeant, Philip, *The Art of Political Storytelling* (London: Bloomsbury Academic, 2020)

Storr, Will, *The Science of Storytelling: Why Stories Make Us Human and How to Tell Them Better* (London: Harper Collins, 2019)

Tsing, Anna, *The Mushroom at the End of the World* (Princeton: Princeton University Press, 2015)

3 The New Commons

Untangling the Enclosure

The American billionaire John Malone, chairman of Liberty Media, surprised the journalists in one of the rare interviews he gave, when he admitted that he has a 'land-buying disease'.[1] He explained this, thus far, undiagnosed phenomenon with his Irish descent, underlining that 'land hunger comes from being denied property ownership for so many generations'.[2]

Today, John Malone is the biggest private landowner in the United States with 2.2 million acres of land across the country. In addition, to honor his ancestry, he bought land worth more than US$300 million in Ireland. Part of his portfolio are several castles, such as Castlemartin in Kilcullen and Humewood Castle in Kiltegan.

John Malone's forefathers were suffering the enclosure by aristocrats in Ireland, migrated to the United States and now, generations later, their ancestor catches the land-buying disease, encloses millions of acres in the United States, and buys the castles of the aristocracy in Ireland. This could be interpreted as an American-style action story: a hero on the intergenerational revenge trip with money as his weapon. Or, it could be interpreted as a sad case of history repeating, with wealthy billionaires in the cast of the 16th-century British aristocracy.

Sadly, most of the people, in case they should suffer from land-buying disease, cannot buy millions of acres in the United States. If David Graeber underlined that you can become king if you can convince everyone that you are legitimate, then John Malone shows that you can also become, some sort of, king if you are able to line up enough dollars.

John Malone's story also sheds light on a factor of wealth accumulation that tends to be overlooked in post-industrial, urbanized Western societies: land ownership.

While the average investor is focused on the global stock market, the avant-garde has already moved to an equity long considered to be outdated: farmland. Whether CNN founder Ted Turner, mining tycoon Gina Rhinehart, or asset managers like Black Rock, the early 21st century is

DOI: 10.4324/9780429288180-3

experiencing a new land rush. But this time, the entry ticket to the game costs a fortune.

Owning land, enclosing land, fencing off land is an age-old obsession of the privileged that has not died out in the 21st century. The biggest player in the landownership game is a long-established institution: the Roman Catholic Church. Its global land possession amounts to 70 million hectares, which is the surface of France and one-half the surface of Spain combined.

The US Land Report states the ownership concentration in the country has increased in the decade between 2007 and 2017: while the 100 largest private landowners owned 27 million acres of land in 2007, ten years later their property increased to 40 million acres. This amounts to the surface of New England. Jacob Waggoner criticizes that landownership in the United States is highly unequal and highly racialized. Waggoner underlines that 'the top 10 (all white) agricultural landowners in the US together own more agricultural land than all racial minorities in the US combined'.[3]

The phenomenon of land ownership concentration is global. Lowder highlights that the 'largest 1 percent of farms in the world operate more than 70 percent of the world's farmland'.[4] According to the World Inequality Database, 'South Asia and Latin America exhibit the highest inequality, with top 10% landowners capturing up to 75% of agricultural land and bottom 50% owning less than 2%'.[5] The official website of the Scottish Government states 'currently 432 private land owners own 50% of the private land in rural Scotland [...] so this means that half of a fundamental resource for the country is owned by 0.008% of the population'.[6]

Landownership concentration thus leads to an increasing amount of people who are excluded form livelihoods, whether in Australia, in Scotland, in the United States, or in South America. In addition, it leads to further enclosure and thus further exclusion. As Winchester brings to mind:

> Central to the concept of owning land is the right to tell others to get off it. One who acquires land gets to enjoy the legally framed Bundle of Rights: the right of possession, the right of control, the right of enjoyment, the right of disposition – and, most relevant here, the right of exclusion.[7]

Enclosure, in its first meaning can thus be described in terms of 'landscape', the change of the ownership landscape through the concentration of property. While John Malone illustrates the contemporary version of (landscape) enclosure, his ancient predecessors suffered from the historic precedent of the same phenomenon.

Historically, concentrated landownership leads to a further drive to enclose land, and land enclosures lead to exclusion of the majority of the population. But as the early enclosure in Great Britain has shown, the protest against enclosure of the land can produce a surprising 'spillover' effect: the opening of the political imagination. Enter Thomas More's *Utopia*.

The origin of modern utopia is a moment of profound transformation of (British) society that started in the 12th century and gradually accelerated in the early 16th century: the local gentry paved the way for excluding a significant part of the rural population of their livelihood by enclosing land. Previously common lands, which were at the disposition of the rural community, were privatized, and that act of theft was later legitimized by the British legislation through the so-called Enclosure Acts. These acts drove a significant part of the countryside population into abject poverty, which partly also resulted in the rise of criminality. As Hertzler explains:

> The increase of pastures by means of the system of enclosure likewise contributed to the volume of thievery. The avaricious gentry and nobility and even 'those holy men the abbots', not only contended with the old rents which their farms yielded, destroyed homes and towns and cultivated lands, thereby dispossessing owners as well as tenants, and enclosed the ground for pasturage that they might lodge their sheep in them and thereby increase their profits.[8]

Land theft from the poor was thus being legalized. The situation of the poor rural population was drastically aggravated through the privatization of commons and the introduction of new methods of agriculture (which reduced the need for labour). The common farmer thus had no more land to cultivate and increasingly fewer options to find employment. The wages decreased, while the profits of the local gentry soared. The rise of unemployment in the countryside led to misery and discontent and, ultimately, paved the ground for migration to the cities. As Simon Winchester points out:

> It was when people began building fences around it, enclosing public land for private purpose, and when others, more cruelly, began dismissing landed tenants and replacing them with livestock-both acts being seen not as social vandalism, but as a means of improving the quality of the land-that matters first started to go awry.[9]

The rise of social inequality that resulted from the enclosure was described by Lewis Mumford in very explicit terms: 'The rich are fattening among the poor; land is gathered in big parcels, at least in England. Extravagant luxury grows on one hand; misery on the other'.[10]

Enclosure thus meant stealing from the commons and excluding a significant part of the population from the possibility to autonomously generate their livelihood.

Obviously, the victims of the enclosure were not able to articulate their voices because they were illiterate and had no political representation. Prior to 1516, the idea of a commonwealth was not explicitly stated, so there was no ideal against which to measure the injustice of the existing society. As J.O. Hertzler observed, before Thomas More 'society was asleep'.[11]

With the publication of *Utopia*, the original title was '*De optimo rei publicae statu deque nova insula Utopia*', Thomas More sent Raphael Hythloday on a journey that left a significant imprint on the course of modern history. Hythloday, the protagonist of More's book, discovers on his journey an island, Utopia, in which he finds the ideal commonwealth which is ruled by an elected monarch. The inhabitants of Utopia do not know private property, there is no class distinction, and everyone on the island is equal in front of the law; the Utopians place a high value on education and religious tolerance and show no interest in gold or money. Hythloday is deeply impressed by the utopian society.

In the context of the social inequalities in England at the time of the writing, *Utopia* has to be understood as anti-enclosure. By establishing Utopia as a faraway island, More situated the ideal commonwealth as a real possibility that could also potentially be realized in the British society of his time. So, utopia, as described by More, is the possibility of an anti-enclosure in the present time. In the epoch in which More lived, other important historical events would shape modernity. As Jameson (2005) underlines:

> the inaugural text of Thomas More [...], almost exactly contemporaneous with most of the innovations that seem to have defined modernity (conquest of the New World, Machiavelli and modern politics, Ariosto and modern literature, Luther and modern consciousness, printing and the modern public sphere).[12]

The utopian anti-enclosure is thus the reaction to a historical transformation which, paradoxically, on one level opens up the world (geographically) while at the same time closing down the possibility for a part of the population to inhabit their own space in this new world. The birth of utopia is also contemporaneous with the birth of modern capitalism which would significantly shape (from a Western point of view) the newly discovered territories. In addition, the new possibilities that would arise through inventions, technological development, and scientific discoveries enhanced the scale of its expansion. The creation of new antagonisms as a result of the extension into all spheres of life profoundly changed the global order. One significant side effect was that the new system would also create new antagonisms. As Lewis Mumford (1922) observes, that 'in setting out for Utopia Sir Thomas More left behind a scene which in political violence and economic maladjustment looks queerly like our own'.[13]

Gradually, the new system and its utilitarian and secular imaginaries also erased the alternatives in the afterlife. Through the decline of religion, which would be a slow process that would accelerate only towards the end of the 19th century, a spiritual vacuum was created that would need to be filled with ideal commonwealths in this world. As Alain Touraine points out, 'the history of utopia began only when society abandoned the image of paradise. Utopia is one of the products of secularization'.[14]

Since the enclosure fenced off the possibility of livelihood for many poor farmers, secularization gradually closed the doors to a paradise in the hereafter for an increasing number of people, and the narrow utilitarianism rang the death bell for imagination, Hytholoday's journey provided a narrative of hope for a different world. This hope not only animated the underprivileged but also those who were looking for an alternative to the reductive drive to the accumulation of (unequally distributed) wealth.

The enclosure can be thus understood on the level of landscape which refers to the concentration of ownership of the land and the ensuing dispossession of the majority of the population. This process can either lead to an enclosure of the mind scape, which refers to the capitulation to the status quo and the reduction of the political horizon, or to a Thomas More style reaction, the enlargement of the political horizon through utopias.

Thomas More's work became a major source of inspiration, and thus opened the mind scape to develop alternatives to the enclosure. Francis Bacon, Tommaso Campanella, and James Harrington followed with their own utopias in the 17th century; the 18th century saw the rise of what Goodwin and Taylor call the 'the liberal-humanitarian brand of utopianism',[15] which accompanied the rise of the bourgeoisie to political power. In the aftermath of the French Revolution, the Utopian Socialists attempted to implement utopian ideas through practical projects, and William Morris integrated utopia in the process of art production. The end of the 19th century saw an explosion in utopian (and later dystopian) fiction. In the mid-20th century, in the aftermath of the Second World War and the horrors of Fascism and Stalinism, utopianism became synonymous with totalitarianism and, thus, suspect.

The mind scape was thus enclosed again: utopian speculation was seen as dangerous because it menaced the status quo in the Cold War. The front (and thereby the political horizons) needed to be clear: the anti-utopian Stalinism against the anti-utopian Capitalism.

The enclosure of the post-World War II mind scape was challenged by Ernst Bloch. In the rather anti-utopian zeitgeist of the 1950s, Bloch published *The Principle of Hope* (1955/1959), a defense of utopian thinking which Bloch saw as anticipatory consciousness. He retorted the reduction of utopian thinking to totalitarian schemes:

> to limit the utopian to the Thomas More variety, or simply to orientate it in that direction, would be like trying to reduce electricity to the amber from which it gets its Greek name and in which it was first noticed. Indeed, the utopian coincides so little with the novel of an ideal state that the whole totality of *philosophy* becomes necessary (a sometimes almost forgotten totality) to do justice to the content of that designated by utopia.[16]

In Bloch's understanding, reality includes already the idea of utopia, and it is impossible to foreclose the Not-Yet-Conscious or the Not-Yet-Become because what is transcendental about history is the desire to transcend. The alluring idea of a hope for the Not-Yet-Become, the realm of political possibilities had a tremendous influence on the 1960s emancipatory movements, which culminated in the student protests of 1968.

Bloch's important contribution was to add to the cold stream of economic conditioning the warm stream of human agency. In his view, utopianism without an understanding of the economic structure is therefore as ineffectual for transcending the existing conditions as pure economic determinism. So economic determinism does not represent the totality of a complex reality. Reality as the dynamic understanding of possibilities is highlighted by Levitas (2005), who underlines that 'reality for Bloch must include the horizon of future possibilities – possibilities which are always plural, and which are dependent on human agency for their actualisation'.[17]

Bloch (1986) argued that the way out of the 'close world that has already become',[18] a world that he considered to be contemplatively idealistic, or merely passive contemplative would be to understand the future as a genuine open process. This kind of open future can not be apprehended by mere contemplation but needs to be approached by 'thinking towards changing the world and informing the desire to change'.[19]

Bloch underlines his positive understanding of utopia with the claim that 'all freedom movements are guided by utopian aspirations'.[20] Underlining the importance of a positive understanding of the future warns that where 'the prospective horizon is omitted, reality only appears as become, as dead'.[21]

How does the prospective horizon look in the 21st century? The previous chapters have outlined the political imaginaries and the stories that are animating the contemporary discussions. In relation to the enclosure of the mind (and the increasing enclosure of land, wealth, and property) the question of the appropriate strategy comes into focus.

The question of how to effect political change, how to undo the enclosure, is animating the political debate (especially on the Left) ever since the French Revolution and is impossible to render in its complexity in this context. But, since the 1960s, the broader lines of the debate run between largely deontological and consequentialist approaches to political ethics. Briefly defined deontology focuses on the question of whether an action is right and when it is good (from a moral point of view); and consequentialism emphasizes that an action is morally good when the consequences of the action are good. On a larger scale, these ethical default settings have also influenced the debate on whether the struggle for alternative political arrangements should prioritize the focus on representation (deontological) or the struggle for a different economic system (consequentialist). These dichotomies are obviously not reflecting the full complexity of the discussion, but for the sake of the discussion on political approaches, they serve as a description of the broad outlines.

The question is which approach is more effective in countering the neoliberal 'war against imagination' and has more potential to develop the political imaginaries that transcend the 'There Is No Alternative' dogma. In essence, the question is about the most effective approach to effect social change to undo the enclosure of property and the mind.

Ruth Levitas clarifies the important difference between political and economic power through an observation by Tom Moylan:

> Moylan reiterates an alarming if increasingly common misconception of Gramsci, claiming that Gramsci identified 'the terrain of ideology and culture as the (rather than a) major site for contesting the dominant power', and contending that under late capitalist conditions economic struggles no longer retain their subversive content.[22]

Through a prioritization of the terrain of ideology and culture, one of the major changing forces in contemporary society, the economy, is left out of the focus of political battles. Subversion has moved into the sphere of lifestyle and representation and is alimented with a strong moral component. Russel Jacoby describes the mindset of what he calls the 'new radical' as utilitarian, liberal, and celebratory with a market-driven approach and diagnoses a conversion or even inversion of leftist thinking. The downsized, fragmented, and market-friendly new radical has thus lost his ideals to change society, and a lack of imaginaries is transcending their respective lifestyle groups. Referencing older utopian ideals, Jacoby points out that: 'Thomas More dreamed of a utopia without war, money, violence, or inequality. Five centuries later, the most imaginative futurists foresee a utopia with war, money, violence, and inequality'.[23] The former co-leader of 'Die Linke' (The Leftist Party), Sarah Wagenknecht, also bemoans that the 'the focus of left-wing politics is no longer on social and political-economic problems, but on questions of lifestyle, consumer habits and moral attitudes'.[24]. She labels these attitudes as 'lifestyle leftism'. In a similar vein, Nancy Fraser describes this attitude of downsizing, fragmentation, and abstraction of the struggle against social injustice as progressive neoliberalism.

The utilitarian, yet moralistic New Radical, Progressive Neoliberal, or Life-Style Leftist has thus retreated into a bubble of emancipation which is limited to their peer group. What is considered to be realism today is to retreat into the private sphere, which is a 21st-century version of utilitarianism as understood by Thomas Babington Macaulay, who famously stated 'An Acre in Middlesex is better than a principality in Utopia',[25] to which Russel Jacoby retorted:

> An acre in Middlesex is better than an estate in Utopia, but is a dreary flat in Middlesex the goal of life? Is it worth surrendering all desires – 'all that about liberty, and so forth' – that go beyond the immediate possibilities? A materialism that values the here and now may be

> beyond reproach, but a utilitarianism that swallows imagination may not be.[26]

Even in utilitarian terms, for the majority of the population, the downsizing of the horizons, the imaginaries, and the expectations results in diminishing returns of the economic conditions. A dreary flat in Middlesex, for instance, on the Yeading Lane in Norholt, is financially out of reach for the largest sections of society. According to the website rightmove.co.uk, a one-bedroom flat in a 1950s red-brick building above a fish and chips joint costs £2,800,000.[27]

If Thomas More's *Utopia* was a reaction to the enclosure of the commons in the 15th century, then the enclosure of the real estate market in the 21st century is a result of the downsizing of the political imaginaries and the ensuing enclosure of the imagination. The absence of the 'principalities of utopia' and the focus on an unimaginative utilitarianism and the private sphere thus makes the acre (or the dreary flat) in Middlesex unreachable for the average citizen. But what good is utilitarianism for, if it does not improve the quality of life and thereby potentially maximize the happiness of the greatest numbers?

In the battle against the enclosure, the extensive focus on either deontological questions and/or the narrow utilitarian perspectives has impaired the debate on opening up the enclosure both on the 'landscape' (the actual concentration of property and the increasing enclosure) and on the 'mind scape' (the enclosure of political alternatives). Another factor is the spatial range of political thinking: the 'global scape'. While it is obvious that many problems of social inequality can only be solved on a global level, the global scape is ignoring the possibilities of entangling the enclosure on a local level.

Is fundamental change thus only possible on the global scape? Is the spatial range of the imaginaries – the scope scape – incorporating the global level, the state level, or the local level? For much of the 20th century, the Marxist idea that the focus of revolutions is the overthrow of state power and the potential endpoint, a world revolution, has influenced the perception of the spatial range. New political projects (and imaginaries) are to target the state level with the perspective of changing the global level. Marx and Engels dismissed local projects by Utopian Socialists as conservative and reactionary. As Paden points out: 'Marx and Engels often noted, the description and construction of these demonstration communities often diverted the utopian socialists from more direct—and possibly more productive—political activities'.[28] From the traditional Marxist point of view, alternative projects with a local range are unproductive experiments which ultimately are escapist. Thus, the Marxist imaginary and its focus on the state have limited the spatial range of political activism.

Latouche's 'decolonization of the imaginary'[29] is thus also an attempt to open up the level of the local for societal experimentation. Varvarousis and Kallis underline the necessity of lived experience as a community when they write: 'Unlike communities of opinion, in communities of practice there is a tendency to bridge, rather than exaggerate differences'.[30]

The quest for alternatives is thus linked to the spatial range. For global phenomena, such as climate change, the strategies are interlinked: effect change in the global political framework and adopt climate-friendly ways of living on a local scale. For the question of social inequality, such a dual strategy is more difficult to articulate. Whereas global policies and policy proposals abound (e.g. Make Poverty History), the multifaceted aspects of inequality make local projects more crucial. Local communities of practice have the potential to create new forms of cohabitation. As Mulgan points out, 'combining imagination and experiment – challenges many dominant political traditions'.[31]

The combination of imagination (or imaginary) and experiment is inherently more feasible on a limited local scale than on a global scale. The imaginary in action which manifests itself in a local community of practice has the potential to create new norms and values, new images and new stories and thus has more potential to generate new imaginaries. Being in common has the potential to create stronger links than having in common, and a community of practice potentially has a stronger connection than a community of opinion; the local level is the only sphere where such an experience is possible in an unmediated way.

So, from a strategical point of view, disentangling the enclosure on the level of landscape, mind scape, and local scape, disentangling the enclosure would mean:

- the creation of common landscapes
- open the discussion on the multiple possibilities of community organization with the intention to put ideas into practice
- focus on the local scope, experimenting with the local conditions

This, is, per definition a process with an open outcome but, without the willingness to experiment on the level of society (on a small scale), the enclosure will continue, and gradually with fewer and fewer resources available, the dispossession and exclusion will accelerate. The land-buying disease among the billionaires is a virulent one, which is not limited to the purchase of land. In order to avoid becoming (or staying) dispossessed and destitute like, for example, the Irishmen who had to flee to the United States in the 19th century, the idea of the commons has to be revitalized. After all, the enclosure of the commons started the vicious cycle in which the new owner of Castlemartin is repeating the same mistakes as the old owner of Castlemartin.

Manifestations of the Commons in the 21st Century

The 10th of December is an important date for anybody interested in the top levels of research in the fields of Physics, Chemistry, Physiology, or Medicine and Economic Sciences. All eyes are on Stockholm and on the Nobel Prize awards ceremony. In 2009, three major subversions of the convention in the prize for Economic Sciences took place. The first two subversions – for the first time, the prize-winner in this field was a woman, and for the first time, the prize-winner was a political scientist – were already groundbreaking. As for the third subversion, the impact was harder to evaluate at first sight. It is connected to the area of research: the Commons, a highly innovative yet, in the mainstream, highly obscure field. The name of the winner is Elinor Ostrom, and her contributions not only to the fields of economic science and political science but also research on the commons were substantial. In her prize lecture, Ostrom summed up what she has leaned in 50 years of field studies and academic research:

> The most important lesson for public policy analysis derived from the intellectual journey I have outlined here is that humans have a more complex motivational structure and more capability to solve social dilemmas than posited in earlier rational-choice theory. Designing institutions to force (or nudge) entirely self-interested individuals to achieve better outcomes has been the major goal posited by policy analysts for governments to accomplish for much of the past half century. Extensive empirical research leads me to argue that instead, a core goal of public policy should be to facilitate the development of institutions that bring out the best in humans.[32]

What Ostrom discovered in her studies, which also included countless field studies in Australia, Bolivia, India, Indonesia, Mexico, Philippines, Poland, and Zimbabwe, was that what she called common pool resources (CPR) were best managed by local communities in an autonomous way with common property settings which are defined by a clear set of rules. Whether forests, fisheries, raw material extraction or irrigation systems, the human and ecological benefits were always significant whenever the state left the local communities to decide on their own resolution mechanisms. Ostrom explains that common pool resources 'can be owned and managed as government property, private property, community property, or owned by no one'.[33] She thus (re-) introduces CPR as a third variable into the well-established dichotomy of private and public ownership, which has dominated the political imagination for the last few centuries (and, in the mainstream perception of property, still dominates).

As an alternative to the narrow focalization on either market or state, Ostrom proposes, in collaboration with her husband Vincent, Charles Tiebout, and Robert Warren, a system of polycentric governance. Ostrom emphasizes that:

> 'Polycentric' connotes many centers of decision making that are formally independent of each other. Whether they actually function

> independently, or instead constitute an interdependent system of relations, is an empirical question in particular cases. To the extent that they take each other into account in competitive relationships, enter into various contractual and cooperative undertakings or have recourse to central mechanisms to resolve conflicts, the various political jurisdictions in a metropolitan area may function in a coherent manner with consistent and predictable patterns of interacting behavior.[34]

For Ostrom, an innovative system of governance is thus based on 'a diversity of institutional settings including human interactions within markets, private firms, families, community organizations, legislatures, and government agencies'.[35] Such an institutional setup would generate higher levels of civic engagement, greater care of common pool resources, such as water, forest, or grazing lands, and thereby produce higher levels of social and environmental sustainability. The commons can thus, under certain conditions, produce superior results than individual or state ownership of resources.

Ostrom calls for a match of institutions to break open the monolithic perception that everything can be state or market regulated and questions the efficacy of large-scale governance:

> Any single, comprehensive set of formal laws intended to govern a large expanse of territory and diverse ecological niches is bound to fail in many of the habitats where it is supposed to be applied.[36]

For Ostrom, the local, self-organized institutions should be a 'a substantial portion of the institutional portfolio of the twenty-first century'[37] in order to tackle the enormous social, economic, and ecological challenges of our era.

It is perhaps no coincidence that Scotland, a country with one of the highest levels of private land-ownership concentration in Europe, was the first country to acknowledge that communities need to be empowered in the acquisition of land. The Land Reform Act 2016 establishes a unique feature into the land right: the prioritization of local communities in the purchase of land. As a consequence, the government, in case a private buyer or a local community compete for the acquisition of land, can enforce the sale to the community bodies. The highest bidder (in most cases, wealthy individuals or funds) is thus not automatically entitled to buy the land, if a local community wishes to establish its right to ownership. The government even financially supports the community claim for ownership with a special fund. The first spectacular case of the implementation of the Land Act was the sale of the island of Ulva. The 2000-hectare island is now owned by the local community who proudly announce on their website:

> The purpose of the community buy-out of Ulva was to secure the future of the island for the islanders and the wider community of Mull, whilst retaining the essence of Ulva that brings back visitors every year and

> inspires so much passion in those who know the Island or have a connection with it through family links.[38]

Currently, the development of the island is on its course: a development manager has been hired, and the 'Community Owned Island' is gradually restoring the derelict buildings, developing plans for a fossil fuel-free existence, and aiming to offer new housing within the next couple of years. Ulva has been inhabited for more than 7000 years, the developments there will provide an example, if the community-based model will secure its existence for the future.

In the south of Europe, in Italy, the citizens demonstrated another strong statement in favour of the commons. In 2011, more than 25 million Italian voters participated in the abrogative referendum on the privatization of the water resources. An overwhelming 95% voted in favour of leaving water as a common resource (and thus against its privatization). Thereby, the Italian citizens directly challenged the privatization policies of their Prime Minister, at the time, Silvio Berlusconi.

Either through the prioritization of local self-organized institutions with government support, like in Scotland, or the protection of common-pool resources against the will of the government, like in Italy, gradually ideas of the commons and polycentric governance become more influential.

Unsurprisingly, given the logic of local self-organization, the biggest impact of the commons can be observed on a regional level. One of the key ideas is community land trust, which is based on co-operative practices in the acquisition of land, similar to the buyout on the Scottish Island of Ulva. In the definition of the Grounded Solutions Network, community land trusts (CLTs) are 'nonprofit organizations governed by a board of CLT residents, community residents and public representatives that provide lasting community assets and shared equity homeownership opportunities for families and communities'.[39] According to the network, there are more than 225 community land trusts in the United States which are all based on the idea of nonprofit community-owned affordable housing.

In 1984, the city of Burlington in Vermont pioneered an approach of public partnership with CLTs. In order to secure affordable housing for low-income and moderate-income families, the city established the Burlington Community Land Trust. The programme was an overwhelming success and created, to date, more than 3000 units of affordable housing. Bernie Sanders, the initiator of the Land Trust, Burlington's mayor at the time, almost became president of the country.

In Italy, 40 years later, the city council of Bologna adopted a new policy that would shape the city-space for years and inspired many Italian cities to follow: the Bologna Regulation for the Care and Regeneration of Urban Commons. In this regulation, the city recognizes the rights of its inhabitants to establish self-organized commons. Through pacts of collaboration, citizens initiatives can run kindergartens, establish agricultural co-ops and neighbourhood markets, decorate public spheres, and develop any project that can improve the city-space. Thus far, more than 90 pacts between the city and community projects have been established. The city of Turin aims its Co-City project to extend the pacts to social-welfare and urban regeneration activities. In 2016, Napoli legally secured the use and regeneration of abandoned city spaces by local communities in an attempt to reestablish the medieval notion of 'Usi Civici'. Another approach towards acknowledging the common good, albeit in economic terms, are the social economies that some regions attempt to promote. The promotion of social enterprise, so nonprofit, solutions to issues such as recycling, childcare, housing, home care, or education is a community-centred approach based on collective control. The Canadian region of Québec has successfully promoted social economies as part of an economic policy based on pluralistic solutions. As one of the main networking institutions for social economy projects, the Chantier, claims:

> Québec's social economy comprises over 11, 200 collective enterprises with overall sales exceeding $47,8 billion, or more than the construction, aeronautics, and mining industries combined. Over 220,000 people work in the social economy every day, in all sectors of activity from retail to the emerging technologies.[40]

Thus, urban commons and co-operative economies are some manifestations to overcome the narrow dichotomy between the state and the private sector on a local level. The partnership of the public sector with emerging civic organizations, or co-operatives, provides an important key for the success of the initiatives.

A lot of initiatives based on the idea of the commons often spring from the grassroots level in times of crisis. As Angelos Varvarousis and Giorgos Kallis explained, in the aftermath of the 2008 Greek credit crisis, a lot of community initiatives came into being in order to provide assistance during the social and economic misery: these included Time Bank of Syntagma Square (where people could barter services), solidarity clinics that stepped in where the national health care system failed, direct producer to consumer networks that organized the distribution of affordable food, solidarity kitchens where people co-produced meals for and with the socially excluded, the People's University of Social Solidarity Economy in Thessaloniki offering free online courses, and solidarity schools which provided free tutorials to school children. Varvarousis and Kallis recognize that these initiatives cannot fully replace a collapsing state infrastructure since they 'they remain marginal, providing for a very small part of the needs of Greek society'.[41] Yet the ongoing existence of such initiatives, like the solidarity clinics, shows the potential of commons-based initiatives.

In 1965, a group of women founded the 'Seikatsu Club' because they were concerned about food safety and environmental damage and began to organize direct producer–consumer networks. The Club turned into a co-operative, with branches in many Japanese regions and one main motto 'autonomous control of our lives'.[42] In a patriachal country, like Japan, this initiative amounts to more than just food. And Seikatsu, with their collective purchase model based on regional autonomy and the ethos of mutual aid became very successful. As Pat Conaty and David Bollier explain, the 'movement brings together over 300,000 consumer and farmer members'.[43] Conaty and Bollier also point to the success of the multi-stakeholder social co-operatives in Brecia, Italy. Since the 1960s families of social services users work alongside caseworkers to develop solidarity co-operatives. The model was widely emulated, and Conaty and Bollier assert that: 'Social co-operatives have spread to all regions of Italy and today number more than 14,000. They provide more than 360,000 paid jobs and more than 40,000 job opportunities for volunteers'.[44]

The Swiss Ministry of Health, in an acknowledgement of the successful social co-operative initiatives, established a Time Bank system in social care. The system, inspired by the initiatives in Brescia is very simple: spend time with pensioners, save it on a time bank and, when the time comes that you are a pensioner, you can use your savings, and someone will take care of you. India is following the lead and attempting to establish a time-bank system for pensioner care nationwide; the state of Madhya Pradesh was the first to establish a time bank in 2019.

The digital commons is another, well publicized phenomena. In this context, it would be impossible to list all the initiatives, but the creative commons, open-source software, wiki, or peer-to-peer production are terms that have left the sphere of programmers and digital content providers and entered the mainstream vocabulary. In an attempt to widen the concept of peer-to-peer (P2P) to the wider sphere of society, Michel Bauwens, Vasilis Kostakis, and Alex Pazaitis have underlined the four main aspects of P2P:

> 1. P2P is a type of *social relations* in human networks, where participants have maximum freedom to connect.
> 2. P2P is also a *technological infrastructure* that makes the generalization and scaling up of such relations possible.
> 3. P2P thus enables a new *mode of production and property*.
> 4. P2P creates the potential for a *transition* to an economy that can be generative towards people and nature.
>
> We believe that these four aspects will profoundly change human society. P2P ideally describes systems in which any human being can contribute to the creation and maintenance of a shared resource while benefiting from it.[45]

The aspects of freedom, the possibility of scalability, the new types of property, and the transition potential are certainly aspects that will have an increasing influence on commons movements. The strategically wise use of technology will decide on whether they are distracting or empowering.

Digital currencies provide a good example of the ambiguity of digital networks. Scalability can be used for speculative purposes or for social purposes. Often the speculative side takes the lead. Fair coins can quickly become just another Bitcoin. However, the potential of digital currencies on a non-scalable local level remains significant.

The commons are not just an innovation factor on the state level, on the local level, on the grassroots level, or in the digital sphere; the wide scope of commons-related projects on a global scale includes ecovillages, co-working initiatives, geo-villages, transition towns, permaculture projects, community gardens, recycling co-ops, time banks of all sorts, insurance and banking co-operatives, irrigation systems, fishing and agriculture commons, and many more initiatives that cannot be explored in the limited framework of this book.

Silke Helfrich emphasizes what makes the commons a special kind of practice: use value is privileged over exchange value. Helfrich asserts that 'when we focus on the commons, we focus on how to shift things from the market sphere to the commons sphere, we focus on how to shift authority and responsibility from state bureaucracies to the many possibilities to *govern the commons* 'by their users'.[46] The 'user' focus defies bureaucracy and, thereby, the 'one size fits all' concept. Due to their citizen focus, the emphasis on reciprocity and co-operation, and the respect for diversity, the commons have the potential to unify disparate social movements. As Helfrich claims:

> The commons concept integrates different worldviews: there are attractors for socialist thinking (e.g. the common possession), for anarchists (the self-organizational driven approach), for conservative thinking (which values the protection of the creation), obviously for communitarian and cosmopolitan ideas (integral, diversity driven approach) and even for liberals (distance to state accountability, respect for individual interests and motivations in joining a community or a project). But it is quite clear that the commons cannot be a single political party programme. That is its strength, and that is why mainstream political players so often misunderstand the commons or even try to co-opt the commons.

Helfrich posits two mottos that frame the open commons approach: 'out-cooperate instead of out-compete' and 'I need the others and the others need me'. And Helfrich also delineates the horizon of the commons: 'local, decentralized *and* horizontal cooperation in distributed networks'.[47]

Despite the obvious success on a global scale, the misunderstanding of the commons is also related to a ghost that haunts a lot of publications on the subject in the 21st century: the ghost of Garret Hardin. In 1968, Hardin published an article with the catchy title 'The Tragedy of the Commons' in which he argued that common resources, pastures open to all, invariably result in overuse and destruction of the commons. The common acceptance of Hardin's thesis put every scholar and every civic initiative on the defensive; they first had to prove that Hardin was wrong before they could articulate their proposals. Even in her Nobel prize lecture, Elenor Ostrom felt the need to refute Hardin's claims. Hardin managed to create a discourse barrier in which the word commons was 'auto-filled' with tragedy. As the many successful commons projects demonstrate, Hardin was wrong. It is not the commons that have led to the overuse and gradual destruction of the natural resources of the planet, but the (privatized) neoliberal exploitation. The real tragedy is the tragedy of neoliberal greed. It is time to put Hardin's ghost to rest.

The commons themselves are not exempted from the neoliberal assault. Especially in the digital sphere the model of peer-to-peer is the foundation of very successful exploitation attempts. As Michael Bauwens observes:

> In terms of knowledge creation, a vast new information commons is being created, which is increasingly out of the control of cognitive capitalism. And the new information infrastructure, cannot be said to 'belong' in any real sense to the vectoralist class. Therefore, my hypothesis is that a new capitalist class is emerging, which I propose to call the netarchists [...]. These are the forces which both 'enable' and exploit the participatory networks arising in the peer to peer era.[48]

While Helfrich underlines the emphasis of use value over exchange value as the core of the commons, the netarchists let the users produce 'use value' (in the sense of the commons) in order to extract the exchange value. Social media platforms, such as Facebook, Instagram, Telegram, TikTok, Snapchat, or Tinder all operate on the principle of extracting the 'commons value' from their users. As an alternative, Bauwens proposes the global commons in which both the use and the exchange value circulate.

In her book, *When Species Meet*, Donna Haraway criticizes that in the theory of value by Marx, 'no companion species, reciprocal inductions, or multispecies epigenetics are in his story'.[49] She emphasizes the encounter value, the 'being with' on a multispecies level. So, adding the concept of encounter value to the existing categories of exchange and use value expands the horizon for the understanding of the commons. However, while the encounter value is a very beneficial category for an enhanced understanding of human interaction, Haraway's post-humanity perspective rather blurs the picture. As long as other species cannot articulate themselves, human beings will always project their ideas onto them and thereby rather distort than clarify the multispecies interaction.

But Haraway's idea of encounter value has resonance in relation to the commons. As Peter Linebaugh emphasizes:

> To speak of the commons as if it were a natural resource is misleading at best and dangerous at worst—the commons is an activity and, if anything, it expresses relationships in society that are inseparable from relations to nature. It might be better to keep the word as a verb, an activity, rather than as a noun, a substantive.[50]

But Linebaugh attaches a warning to the idea of commoning. In the focus on encounter value and regenerative practice, communing should not play the game of the capitalists who want to privatize wealth and socialize poverty.

So, the focus on use value and on encounter value make the commons a very attractive alternative to a system largely focused on exchange value as its only norm. But as for the question of how to organize commons and what is their most essential prerequisite a historian like Peter Linebaugh comes to one conclusion: the land.

Through the example of Matsutake foraging and trade, Anna Tsing, is illustrating how survival is possible without landownership. The precarious existence of the foragers is recompensed with a sense of freedom and a, sometimes fleeting, sense of community. The social relationships among the foragers, if they are not part of the same extended families, are unstable, full of unpredictable encounters with other humans and nature, and the forest in which they find their livelihood is not their property. In that regard, the Matsutake is part of the latent commons, a loophole of potential in a system that has foreclosed the options for economic survival for many people. Tsing describes her observations as follows:

> I search for fugitive moments of entanglement in the midst of institutionalized alienation. There are sites in which to seek allies. One might think of them as latent commons. They are latent in two senses: first, while ubiquitous, we rarely notice them, and, second, they are undeveloped. They bubble with unrealized possibilities; they are elusive…[51]

In Tsing's interpretation, latent commons are not good for everybody since every 'instance of collaboration makes room for some and leaves out others'[52]; they do not institutionalize well due to their volatility, and they are not redemptive since their focus is on the here and now. Most importantly, Tsing emphasizes, 'humans are never fully in control'.[53] It would have a been highly interesting encounter if Tsing, during her research on Matsutake in Japan, had met the courageous women of the Seikatsu co-operative. For more than 55 years, their project demonstrates that commons have more potential than the precarious existence of the Matsutake foragers.

The latent aspect of commons is certainly an important aspect in the beginning; the search for sites in which to find allies are important starting points and Tsing's encouragement to notice these latent commons is an important step towards developing possibilities. Tsing notices the possibility to explore the interstices and explains:

> Even as entrepreneurs concentrate their private wealth through building alienation into commodities, they continue to draw from unrecognized entanglements. The thrill of private ownership is the fruit of an underground common.[54]

One of the more militant ways to create latent commons is the French ZAD movement. ZAD stands for 'Zone à défendre', so, zone to defend. The approach of the ZAD is not new: occupying land that is in danger of destruction whether through expansion of nuclear plants, the construction of airports, motorways, dams, or housing. The Zadists squatted endangered zones, sometimes for several years, for example, in Meuse, Averyron, Seine Maritime, or in Notre-Dame-des-Landes in order to stop 'developments' that accelerate commercialization and environmental destruction. The prospective horizon of these latent commons, settlements of protesters, is open and highly precarious. The end of each ZAD occupation is marked either by the abandonment of the initial project by the investors (cancelled airport, dam, or amusement park projects) or the eviction of the Zadistes by the local authorities. In some cases, the commons are thus protected through the latent commons of the ZAD movement.

One of the ongoing ZAD projects is in Germany, in the Rhenanian village of Lützerath. The aim of the Zadistes, who started to occupy the space in 2020, is the protection of the village against its looming destruction by the energy company RWE. In the Garzweiler project, RWE plans to transform the whole region, which also includes five more villages, into a gigantic coal mine. Parts of the village were already destroyed by caterpillars and now the activist squad the village and organize protests. In March 2022, the German courts ruled in favour of RWE, but the protesters were planning to resist the eviction and the destruction of Lützerath. In January 2023 the police evacuated all the protesters and thereby terminated the experiment of the ZADist movement and their transitory common. The sad irony of Lützerath is that it demonstrates the reverse logic to Tsing's underground of the commons: it is the privatization of what is under Lützerath that leads to the destruction of the village.

In their description of the 2008 Syntagma Square protests in Athens, Varvarousis and Kallis describe the moment of the Greek anti-austerity protests when the community of the commoners constituted itself. The Square became a locus of a wide array of disillusioned Greek citizens, each with their own story, their own agenda, and their own political worldviews. Through the characteristic slogan 'Make your statement! Find your room in Syntagma Square and express yourself', the organizers of the protest created an open forum that encouraged everyone. The whole spectrum of Greek society was thus taking the microphone, teenagers and pensioners, Greeks and refugees, left-wing political activists and right-wing populists, white-collar workers and punks. The inclusion of a large spectrum of opinion thus made the formation of one group with specific norms, values, and stories impossible. Vavarousis and Kallis identify this as an example of a liminal commons:

> In a liminal commons, the glue that brings the actors together is the practical production of the common. A collective identity is neither a precondition nor the purpose of the process and is discouraged when it puts obstacles in the way of common production.[55]

The value of these liminal commons, comparable to Tsing's description of latent commons, is in the practical collaboration of members with different value systems and identities. On Syntagma Square, some of the slogans during the protest were reflecting the ground zero of identity in times of a collective crisis. According to Varvarousis and Kallis, placards declared slogans such as 'We are nobody' or 'We don't know where we are going but we are not going back'.

Latent commons and liminal commons are certainly important moments in the story of the battle, the initial moment of the constitution of the protest. But these stories quickly evaporate without bigger consequences, as the Greek example of the conversion of the Syriza or the Spanish example of the conversion of the Indignados have demonstrated. For the battle of the story, liminal protagonists are too elusive for identification.

Despite the fact that Elinor Ostrom's CPR approach has been criticized for putting too much emphasis on a strong group and so is situated on the opposite end of the liminal commons, her operating rules for successful commons are still relevant. Ostrom posits seven broad types of rules for organizing sustainable commons:

1. boundary rules that specify how actors were to be chosen to enter or leave these positions
2. position rules that specify a set of positions and how many actors hold each one
3. choice rules that specify which actions are assigned to an actor in a position
4. information rules that specify channels of communication among actors and what information must, may, or must not be shared
5. scope rules that specify the outcomes that could be affected
6. aggregation rules (such as majority or unanimity rules) that specify how the decisions of actors at a node were to be mapped to intermediate or final outcomes
7. payoff rules that specify how benefits and costs were to be distributed to actors in positions[56]

Commons that are defining borders, defining the roles of the participants, and establishing a system of governance of sorts are thus the minimum organizational requirement. It is important to note here that a common interest and the agreement on common rules are not the same as a coherent common worldview. So, Ostrom's operating rules are not restricting the potential of diversity in the community. But, without encounter value or without a spirit of commoning, these minimum requirements are not enough to sustain the community. In the long-term perspective, it is still easier to separate groups than to unite them. In order to resist the 'divide et vince' of the neoliberal paradigm, latent or liminal commons with a precarious group identity are not sufficient. But, as Ostrom points out, some commons, such as the northern Norwegian fishing commons, were established in the 11th century and remained unchanged until 1993. The problems with the Norwegian fishing commons started with the introduction of industrialized fishing practices and ownership concentration. So, it is the neoliberal scalability that is now endangering not only the livelihood of the small fishermen but also of the balance of the submarine ecosystem. The process of enclosure is thus ongoing, as Neera Singh emphasizes: 'As enclosure of the commons intensifies, and all realms of life are increasingly commodified, the calls for reclaiming commons or inventing new commons are growing strong'.[57]

The contemporary dangers for the commons are thus not only the 'old school' privatization but also the scalability of previously (small scale-) balanced activities, the fragmented identities and 'story of the battle' strategies of latent commons, and the lack of horizontal networking. As the examples of the Italian urban commons in Milan, Turin, and Naples have shown, collaboration with local municipalities or, in the Ulva case, even with the state show a greater promise of sustainability. To simplify, there are three alternatives for the commons movement: collaboration, co-optation, or isolation. The current crisis context and the potential of digital technologies can have either a harmful or empowering aspect for new projects. The quest for the site to seek allies is linked to the discovery of spaces of possibility that allow both freedom and collaboration.

The Lessons from Intentional Communities

Visitors who went to the only hotel in New Harmony in the year 1826, did not come to enjoy the view of the wild nature of Indiana or the beauty of the Wabash River. Through the windows of their rooms, they could see the opposite of untarnished nature: the most spectacular experiment on human civilization of its time. The town of 800 people bustled with activity, and the men and women on the street were driven as if by an unknown spirit. Upon closer observation, it became clear that they were not religious but that something else caused this upheaval of human activity. When the school bell rang, boys and girls went running into their classrooms. With 200 school-age children, the town was a young and energetic place. At six o'clock in the evening, something highly unusual happened: people went home after eight hours of work. The public square was filled with workers, farmers, and artisans in active discussions or just enjoying the evening sun. Some entered the local convenience store to do their shopping, but nobody paid for their tobacco or their hairbrushes. Dollars were a useless currency in this town. On a large table, outside due to the unusually good weather, people were serving food. Everyone picked a plate and joined the meal. No slaves – harvesting or serving food – can be seen anywhere. Unlike in neighbouring provinces, slavery is forbidden, and everyone is considered equal and free. On that Friday evening, the city all of a sudden became agitated, everyone rushed out of their houses to the main square, faces were filled with an air of anticipation, the man who made all of this happen was about to come back after a long journey abroad. Even the hotel visitors got out of their slumber and rushed to the main square; after all, they wanted to see the visionary who made this unreal place happen. The carriage arrived, and the people's voices became more excited, as everybody wanted to talk to the man, ask questions, get ideas, and make suggestions. The man, who stepped out of the carriage, looked tired from his long journey from England but managed a smile before he started to shake hands, exchange a couple of words here, give a couple of claps on the shoulder there, and waved at people there. Robert Owen was back in New Harmony.

New Harmony was a shift in a phenomenon that started already thousands of years earlier in what is today Italy: a group of people separated itself from the mainstream of society in order to experiment with a different way of organizing their lives. The first of such intentional communities on record was initiated by Pythagorus around 600 BC. His ideas for Homakoeion were straightforward: the commune was to be vegan and based on the values of knowledge, intellectual discourse, and equality of the sexes. An element of mysticism was added to the mix that could be considered either surprisingly contemporary or based on archetypical needs for equality, community, and new ideas.

For more than a thousand years after Homakoeion, at least in the occidental sphere of the globe during and after the rise of Christianity, the idea of intentional community was dominated by largely monastic groups that were first underground and later organized by the Church. In the 16th century after the Reformation, Protestant grassroots initiatives started to dominate the foundation of new intentional communities. In combination with a resettlement to the United States and the opportunities to acquire communal territory, groups like the Hutterites, the Bruderhof, and the Amish established communes that exist until the present day.

The French Revolution was a seismic shift not only in the political landscape but also in relation to intentional communities, since it demonstrated the possibility of altering the social order in a secular way. At the same time, the resulting disillusion in the aftermath of the Napoleonic wars led to a critical reevaluation of the idea of state revolution. As a reaction in the 19th century, the utopian socialists enter the historical stage and shift the focus of political struggle. Abbe Moreilly, Francois Noel Babeuf, Count Henry de Saint-Simon, Charles Fourier, Etienne Cabet, Louis Blanc, and Robert Owen are key figures. According to Hertzler:

> The Utopian Socialists and their immediate followers were all men in deadly earnest. While they were dreamers, possessed by vague, intangible, but wonderfully broad-horizoned ideals of humanity's perfection, they were sincere in the extreme, and were resolved to make the dreaming come true.[58]

What distinguished the Utopian Socialists from the early modern Utopians is the fact that some of them translated the utopian visions into practice which resulted in intentional communities that attempted to implement a new form of social coexistence. For Fourier, Saint-Simon, and Owen, the social experimenters among the Utopian Socialists, the focus of political rearrangement was not the state but the small community. Goodwin and Taylor (1982) see this line of Utopian Socialism as a continuation of an old tradition:

> The image of utopia as a relatively small island or enclave jealously protected from contamination by the external world has persisted throughout the history of Western civilization. The city states of Ancient Greece and Renaissance Italy, the medieval monasteries, the lost islands discovered by the great explorers of the fifteenth and sixteenth century, the Fourierist Phalanxes and Owenite colonies of the early nine-teenth century, the 'city beautiful' and town planning traditions of the late 1800s, the communal movements of the 1960s – all these traditions point in the direction of a utopianism based on small-scale communitarianism rather than the large, expansionist nation-states and empires.[59]

The focus on small-scale communitarianism thus became a popular emancipatory strategy when the revolution on a bigger scale did not transform the society. Which brings us back to New Harmony and Robert Owen.

Owen, a Welsh with a working-class background, had already successfully established the New Lanark cotton mill in Scotland and demonstrated his skills not only as a visionary but also as a successful manager of a company of more than 2000 employees. In 1825, when he bought the town of Harmonie (named after the Harmonite settlers who left the place and sold it to Owen for the, then, considerable sum of \$150,000) the hopes were high due to Owen's successful New Lanark project. Harmonie became New Harmony, and within a short time hundreds of people followed Owen's invitation to join the community of equality. When Owen came back from one of his many journeys in 1826, it was clear that his social experiment faced some structural problems. Apart from the absence of leadership (due to the absence of the uncontested leader Owen), the influx of people created housing problems, too few of the new community members had experience in agriculture, and there was a lack of skilled labour in some of the workshops which led to the interruption of the industrial production. In addition, indiscriminate access to New Harmony also attracted a fair share of 'free riders' who just profited from the work of others without contributing to the community. As a result, the deficits started to grow and the dissatisfaction along with it. Despite its remarkable achievements in free education for everyone, humane work schedules, equal treatment of men and women, abolishment of slavery, and money free economy, the New Harmony experiment ended in 1827. As Yaacov Oved observed: 'The only prosperous venture was the local hotel, where the many tourists and the curious who came to see with their own eyes Robert Owen's famous experiment were put up'.[60] The tourists interested in social experiments thus had to find other places to go to.

Despite the failure of New Harmony, the spirit of Utopian Socialist experimentation gave birth to many intentional communities. Barthélemy Prosper Enfantin gathered thousands of people in Paris around his interpretation of Saint Simonism. Charles Fourier developed ideas for communal cohabitation spaces called phalanstères (the neologism is based on phalanx and monastery, thereby revealing its partial religious inspiration). Equality, also between the sexes, income sharing, and cohabitation (in a space where Fourier meticulously fixed all the architectural details, including the exact number of inhabitants: 1620) were practiced. A number of intentional communities in the United States were founded, inspired by Fourier's ideas (30 Fourierist associations existed around the 1840s). In the end Fourier so profoundly believed in the feasibility of his projects and in the accuracy of his analysis of human nature, that he rejected the label 'utopianism' in favour of 'science'. As Levitas (2011) remarks, quoting Fourier:

> What is Utopia? It is the dream of well-being without the means of execution, without an effective method. Thus, all philosophical sciences are Utopias for they have always led people to the very opposite of the state of well-being they promised them.[61]

In the same vein as Fourier, Saint-Simon also claimed a scientific basis for his social experiments, as Levitas (2011) explains: 'Saint Simon's ideal of society consisted of three classes, each appropriate to the three human types: scientists, artists and producers; this triadic view of human nature was claimed to be scientific'.[62] This also indicates that the priorities in most utopian socialist projects were similar to their early modernist antecessors: education, knowledge, and art were highly valued, but now they were given pseudo-scientific justification.

In the aftermath of the 1848 revolution in France and the publication of the *Communist Manifesto*, the development of utopian alternatives reached a new level of dynamism, as the list of a post 1880 publication that Levitas compiled demonstrates:

> One can identify a list of texts that are 'classics' of utopianism, of feminism, of sociology, all written within a short period of each other: 1888 Edward Bellamy *Looking Backward* 1890 William Morris *News from Nowhere* 1893 Emile Durkheim *The Division of Labour in Society* 1895 Friedrich Engels *The Origins of the Family, Private Property and the State* 1897 Edward Bellamy *Equality* 1898 Charlotte Perkins Gilman *Woman and Economics* 1905 H. G. Wells *A Modern Utopia* 1911 Charlotte Perkins Gilman *Herland*[63]

For Mumford and Hertzler these late 19th-century utopias were already a dilution of the initial utopian spirit. Hertzler calls them 'pseudo-utopias' because 'with the perfecting of theories of history and the growth of the idea of development or evolution, real Utopias ceased to appear'.[64] The utopian experimentation in the form of intentional communities did not produce any new impulses; instead, some part of the utopian energy shifted into the world of ideas. But not all the authors were satisfied with merely writing about new forms of society.

Among the influential utopian thinkers at the end of the 19th century was William Morris. The question of the perverting influence of machinery was an important aspect of his reflections on a better society. Morris had a significant impact on the Arts and Crafts movement, which was the first major attempt to implement utopian ideas in the art world.

The Arts and Crafts Movement, and Morris' significant contribution to it, was not just another fashion in the art cycle (although some works were highly valued by art aficionados of the time) but had some deep implications for the role of art and the process of making art. Questions about the role of technology in the process of art-making, the division of labour, possible alternative social arrangements, and the role of art in facilitating a new political understanding were treated as having equal importance with questions about aesthetics.

As Great Britain was at the forefront of a brutal industrialization process, social inequality became a more pressing issue. At the core of the problem, one of the key inspirations for Morris, John Ruskin, located the problematic in the division of labour. As Ruskin (1853) argued:

> we want one man to be always thinking, and another to be always working, and we call one a gentleman, and the other an operative; whereas the workman ought often to be thinking, and the thinker often to be working, and both should be gentlemen, in the best sense. As it is, we make both ungentle, the one envying, the other despising, his brother; and the mass of society is made of morbid thinkers and miserable workers.[65]

The outcome of this division of society was not just material inequality but also spiritual impoverishment on both sides of the divide. For Morris, who was inspired by craft-based art-making during the Middle Ages, the question of pleasure in labour became a key concern. In *Art and Socialism*, initially published in 1884, Morris criticized the modern division of labour that separated the craftsmen from the artist, which had negative consequences for both, for the 'artist came out from the handicraftsmen, and left them without hope of elevation, while he himself was left without the help of intelligent, industrious sympathy'.[66] For Morris, socialism and art are profoundly linked because art disconnected from society, and only geared towards commerce (or in Morris' words the 'greed of money'[67]), becomes purposeless.

Morris explicitly wrote his most famous work *News from Nowhere* (1890) as a reaction to Edward Bellamy's influential book *Looking Backward* (1888). Morris' world in *News from Nowhere* is a deurbanized London as opposed to the technology-driven industrialist projections by Edward Bellamy. Both looked at how the world should be in the year 2000. In the retrospective, Bellamy's vision, his technology-driven utopia has become reality. In the light of the permanent crisis, the question is whether it is a good society. Morris' visions, on the other hand, are a source of inspiration for many intentional communities who aim for a society that prioritizes humans over technology.

In the same vein as Morris, Lewis Mumford criticized the dependence of late modern utopias on technology. For Mumford, 'these utopias are all machinery: the means has become the end, and the genuine problem of the ends has been forgotten'.[68]

The critical attitude towards technology also manifested itself in a (post-) romantic idealization of the purity of nature. One of the most prolific figures to combine the return to nature with an attitude of civil disobedience and retreat from civilization was Henri David Thoreau. A lot of his writings became poster quotes for intentional communities that advocated for environmentalism and ecologic regeneration. Thoreau, for instance, insisted that:

> We must learn to reawaken and keep ourselves awake, not by mechanical aids, but by an infinite expectation of the dawn, which does not forsake us even in our soundest sleep. I know of no more encouraging fact than the unquestionable ability of man to elevate his life by a conscious endeavour. It is something to be able to paint a particular picture, or to carve a statue, and so to make a few objects beautiful; but it is far more glorious to carve and paint the very atmosphere and medium through which we look, which morally we can do. To affect the quality of the day, that is the highest of arts.[69]

The essential features for rural secular intentional communities were thus established: small-scale communitarianism, egalitarianism, experimental approaches to human cohabitation and co-working, valorization of art and craft, and environmentalism. The 1960s and the rise of the counterculture movement opened the floodgates for intentional communities. While very few of the hippie communes lasted, some do still exist today. For instance, Twin Oaks and Eastwind Community. The principles for Eastwind Community and Twin Oaks are explained by the Federation of Egalitarian Communities (FEC) of which both are members. According to FEC, each of its members:

- Holds its land, labor, income, and other resources in common.
- Assumes responsibility for the needs of its members, receiving the products of their labor and distributing these and all other goods equally, or according to need.
- Practices non-violence.
- Uses a form of decision making in which members have an equal opportunity to participate, either through consensus, direct vote, or right of appeal or overrule.
- Actively works to establish the equality of all people and does not permit discrimination on the basis of race, class, creed, ethnic origin, age, sex, sexual orientation, or gender identity.
- Acts to conserve natural resources for present and future generations while striving to continually improve ecological awareness and practice.
- Creates processes for group communication and participation and provides an environment which supports people's development.[70]

Since 1967, Twin Oaks, which own 400 acres in Virginia and currently has 85 adult members, practices these principles. According to FEC, the community generates their income through tofu production, book indexing, seed growing, cheese making, biological agriculture, and hammock making. East Wind Community was funded in 1974, and currently 60 adults are living there on 1045 acres in Missouri. The business activities include biological agriculture, nut butters, and Utopian rope sandal making.

The majority of the communities in the FEC have less than ten members, such as Sandhill Farm and Oran Mór in Missouri, Eastbrook Community Farm in Walton, Le Manoir in Quebec, and Open Circle in Virginia.

The FEC types of communities are part of what the umbrella organization Fellowship for Intentional Communities (FIC) would label as communes that are 100% income-sharing. In addition to communes, the FIC lists ecovillages, cohousing, co-ops, and religious communities as part of their network. For the sake of focus, this chapter only discusses secular communities.

In relation to the intentional communities in rural areas, Adrian Shirk observes a tendency of a return to the roots of Native Americans, despite the fact that the participants are anything but Native American. Shirk points out:

> we're talking almost exclusively about white Protestants, or people who have at least gone rogue from a European Protestant theology. And it seems to me that the life those communities created, which we call American utopias, was more or less a paltry mimesis—consciously or not—of the kind of life that North American Indigenous people had been living on that land for centuries and were, at that time in the mid-1800s, defending with their blood and bodies.[71]

Shirk observes that utopia-making 'emerges in force especially during times of economic and social precarity'.[72] The precarity and the estrangement from mainstream society are thus the driving forces into seclusion.

Lucy Sargisson, describes the process of estrangement as a physical estrangement, the separation of the community from the rest of society in order to focus on rebuilding the estranged self. Often this challenges the participants because not only have they come to terms with often harsh living conditions in the secluded rural areas but they also have to balance out their relationship with the group. Sargisson observes that, based on her observation of over fifty intentional communities in Britain and New Zealand, the most enduring challenge for intentional communities 'is the need to balance the competing needs of individuals and the group'.[73] Since the most decisive factor is group cohesion and a collective vision the group formation process is the most problematic in communities that place a very strong value on social coherence. The risk of these procedures is, according to Sargisson, that this estrangement 'places community members in a position of permanent social otherness'.[74]

While it might seem comparatively easy to reconnect with nature, the flaws of social conditions that each member brings in from the 'outside' society are harder to cope with in small-scale communities without family bonds and traditions while living together in the periphery. As Sargisson clarifies:

> intentional communities are at once inside and outside the wider community. They are embedded: they stem from it. They are part of it. However, they exist because they are deeply critical of dominant norms, values and practices. It follows that attitudes to 'the outside' will be critical.[75]

Geoph Kozeny has observed the same interrelatedness between outside societies and intentional communities. Kozeney underlines that:

> Communities draw their membership from society at large, and those members bring with them generations of social conditioning. The attitudes, behaviors, and institutions prevalent in the broader society–including the very things we seek alternatives to–are a significant part of our upbringing. Merely identifying a problem and expressing a

> desire to overcome it does not mean that we presently have the perspective or skills needed to transcend it. The problems we see 'out there' in the mainstream–greed, dishonesty, excessive ego, lack of self-esteem, factionalism inadequate resources, poor communication skills, you name it–all manage to find a significant role in alternative cultures as well.[76]

Alexa Clay sees a potential mistargeting for intentional communities, since the need of belonging can often be exploited by the wrong people. Clay sees as a potential danger that 'dreamers drifters and seekers in need of belonging, the needy and wounded, and the egomaniacal and power-thirsty are a dangerous constellation of actors for sustaining a community'.[77] To avoid this mix is most responsive to an invitation to join an intentional is a decisive challenge. Many communities have addressed this problem by establishing trial periods for prospective members. For example, in Twin Oaks, potential members have to visit the community for three weeks before being considered for provisional membership. Then they have to leave for a month, so that the members can deliberate on their case. In case of acceptance, there is a six-month provisional membership, and only in the case of a positive verdict by the community can they sign a membership agreement.

Even if the functioning communities have created what appears to be a modus operandi, especially in the commune-type community, the question prevails whether it is possible to change human conditioning in the framework of a rural, high-intensity group.

In addition to the 'human conditioning factor', often intentional communities fail for very mundane reasons, such as the same finance-driven economy that they try to escape from is catching up with them. Robert Owen's New Harmony experience is just one of many economic failures of communities. But despite the failed projects, many communities have successfully established themselves. As Bill Metcalf emphasizes there are 'multi-million-dollar operations, such as Findhorn Foundation (UK), Damanhur (Italy), and Twin Oaks (United States)'.[78]While Alexa Clay relativizes the economic success of a longstanding project like Findhorn Foundation (the community was founded in 1972 and is one of the showcase projects in the UK) by showing that the institution barely breaks even (an income of £2,393,542 facing an expenditure of £2,350,411), she concedes that the judgement of the economic performance is not adequate when compared to the business world. Clay compares the intentional communities to start-ups which have a failure rate of 90%. Intentional communities also don't fare badly in comparison to Fortune 500 companies. As Clay demonstrates, out of the Fortune 500 companies listed in 1955, more than 80% are gone, and S&P companies have an average life span of 15 years. In comparison to the rather poor longevity performance, Clay asks: 'Can we really expect more longevity from experimental communities?'[79] Placing communities into a larger organizational context, Clay concludes that:

> the more relevant drivers that cause many communities to unravel sound more like the challenges afflicting any organization today: capital constraints, burn-out, conflict over private property and resource management, poor systems of conflict mediation, factionalism, founder problems, reputation management, skills shortage, and failure to attract new talent or entice subsequent generations.[80]

In that understanding, intentional communities should be considered society start-ups. And if start-ups that are dealing with only one aspect of the society, the economy, have a failure rate of 90%, how high is the permissible failure rate of start-ups that are dealing with establishing a whole society anew? After all, as Lucy Sargisson emphasizes, these experimental communities are 'provoking a key utopian function: fresh perception of the limits of the possible'.[81]

As the examples of CLTs and the urban commons approach of, for example, Bologna have shown the urban cohousing and cohabitation movements are increasingly popular and often supported by local institutions. Urban spaces are per definition not well suited for isolation and closed communities so the approach to intentional communities is in general more practical oriented and less focused on reconditioning humans or the establishment of ultra-tight group ties. There is also a perceptible generational shift, whereas the urban communes are largely baby boomer projects, many of the urban co-living projects are organized by the more technology-attuned millennials.

Driven by the search for a new mission, Jessy Kate Schingler left her job at the NASA where she worked in the Nebula team for cloud computing. She moved to San Francisco, turned her 7500 square -foot mansion into a creative residence and founded the cohabitation space Embassy Network. The idea is simple: people all over the world connect through a network of co-living spaces. Schingler (who now works for the Open Lunar Foundation on projects of governance for possible moon explorations), created a variety of self-governing collectives and place-based intentional communities like District Commons, Open Door or the Embassy Network. When interviewed by a local newspaper, Schingler explained: 'What I liked about space was the idea of creating human settlements. But now I found out that I don't have to go to space to do that'.[82]

The Embassy Network addresses young urban professionals. On the offer are mansions, lofts, converted hotels, and apartments not only in the fashionable districts of cities like Barcelona, Berlin, or San Francisco but also in rural places like Costa Rica or Puerto Rico. The cohabitation is not limited to sharing living spaces, meals, and cars but also explicitly appeals to the idea of professional networking, peer support, and knowledge exchange. The aim is to create a 'network of place-based communities experimenting with new forms of governance and solidarity'.[83]

The project is part of the Sharing City Movement which includes cohabitation projects to support the aged like Humanitas or Babayagas House, nonprofit car-sharing initiatives like Ride Austin, or cohousing consulting projects like Mietshäuser Syndikat. The Sharing City movement is collaborating successfully with mayors, for example, in Seoul. Neal Gorenflo, the co-founder of the movement emphasizes that the focus is on 'practical, interpersonal action-sharing – that ordinary citizens can engage in and help each other and the city as a whole'.[84] The initiatives in Sharing City Seoul focus on citizen stakeholder initiatives that address problems, like pollution or social isolation, and won the city international prize for sustainable development. Now the London-based SharingCities.eu is promoting the concept to European cities and networks, like the Sharing City Alliance and the Paris-based OuiShare, which are developing new forms of local sharing initiatives. All of these initiatives are based on technology-driven networks used in order to impact the specific local conditions. The language is also tech-inspired: peer-to-peer, networks, nodes, catalysts, and (problem) hacking. But maybe, to add another tech-inspired term, a better framing of intentional communities would be 'society start-ups'.

The tech executives who are sharing 7500 square-foot mansions and countryside hideouts in Costa Rica could be easily dismissed as 'hipsters', in the same way as the residents of East Wind Community who are dancing in colorful clothes around a tree could be laughed off as 'hippies'. But the new sharing movement is addressing audiences that could never be reached by the traditional rural-based intentional communities. And the differences in lifestyle, language, and age should not be leading to a division between emancipatory movements. The 'devide et impera' tactics of neoliberalism can only be overcome if there are more diverse 'styles' of cohabitation and intentional communities on display. The 'hipsters' and the 'hippies' are increasingly joined by the middle class in projects like Serenbe, and the overall idea is to create a multitude of society start-ups. After all, one of the main functions of intentional communities is not only to create better living conditions on a local level but also to generate a diversity of templates for a better living. East Wind Community is part of the same mosaic as Embassy Network. The picture only makes sense if people from all ranges of life find meaning in it. A variety of images and imaginaries and open networking (as opposed to the recreation of 'enclosure' in the forms of narrow lifestyles) are therefore an essential precondition. In his book *Places and Placelessness*, Edward Relph describes the biggest danger for any society start-up: the transformation of a place to a non-place. Relph explains:

> Placelessness describes both an environment without significant places and the underlying attitude which does not acknowledge significance in places. It reaches back into the deepest levels of place, cutting roots, eroding symbols, replacing diversity with uniformity and experiential order with conceptual order.[85]

A too strong focus on uniformity and conceptual order would convert the eutopian impulse to create a good society, into an outopian reality:

the non-place that was supposed to be turned into a good place becomes placeless.

Lucy Sargisson and Lyman Tower Sargent underlined one of the most important challenges for society start-ups: 'People who live in intentional communities need to learn how to balance their own needs with those of the groups'.[86]

Place matters, people matter, but the self also matters; this could serve as a basic formula for society start-ups.

Notes

1 Monte Burke, 'John Malone Overtakes Ted Turner as Largest Individual Landowner in the U.S.', *Forbes*, 10 March (2011) <https://www.forbes.com/sites/monteburke/2011/03/10/john-malone-overtakes-ted-turner-as-largest-individual-landowner-in-the-u-s/> [accessed 5 January 2022]

2 Robert Frank, 'John Malone Now Biggest Landowner in the U.S.', *Wall Street Journal*, 12 October (2011) <https://www.wsj.com/articles/BL-WHB-4892> [accessed 5 January 2022]

3 Jacob Waggoner, 'Land of the Free or Land of the Few? Local Concentration in US Landownership' (2020), *Harvard Mellon Urban Initiative* <https://mellonurbanism.harvard.edu/rules-engagement-forming-guidelines-design-schools-working-marginalized-communities> [accessed 5 January 2022]

4 Sarah K. Lowder, Marco V. Sánchez, and Raffaele Bertini, *Farms, Family Farms, Farmland Distribution and Farm Labour: What Do We Know Today?*, Agricultural Development Economics Working Paper 19-08, Rome, FAO (2019), p. v.

5 Luis Bauluz, Yajna Govind, and Filip Novokmet, 'Global Land Inequality, World Inequality Database' (2020) <https://wid.world/news-article/global-land-inequality/> [accessed 10 January 2022]

6 *The Land of Scotland and the Common Good: Report* (2014), Land Reform Review Group <https://www.gov.scot/publications/land-reform-review-group-final-report-land-scotland-common-good/pages/61/> [accessed 10 January 2022]

7 Simon Winchester, *Land* (London: William Collins, 2021), p. 215.

8 Joyce O. Herzler, *The History of Utopian Thought* (London: George Allen & Unwin, 2012, original work published 1923), p. 131.

9 Winchester, p. 168.

10 Lewis Mumford, *The Story of Utopias* (New York: Boni and Liveright, 2008, original work published 1922), p. 39.

11 Herzler, p. 121.

12 Fredric Jameson, *Archaeologies of the Future: The Desire Called Utopia and Other Science Fictions* (London: Verso, 2005), p. 1.

13 Mumford, p. 38.

14 Alain Touraine, *Can We Live Together? Equality and Difference* (Stanford: Stanford University Press, 2000), p. 29.

15 Barbara Goodwin and Keith Taylor, *The Politics of Utopia: A Study in Theory and Practice* (London: Hutchinson & Co, 1981), p. 176.

16 Ernst Bloch, *The Principle of Hope* (reprint ed. 1995), trans. by N. Plaice, S. Plaice, and P. Knight (Cambridge, MA: MIT Press, Vol. 3:3, 1986), p. 15.

17 Ruth Levitas, *The Imaginary Reconstitution of Society: Or Why Sociologists and Others Should Take Utopia More Seriously*, Inaugural Lecture, University

of Bristol, 24.10.2005 <http://www.bristol.ac.uk/depts/sociology/staff/pubs/levitasinaugural> [accessed 12 February 2019], p. 7.
18 Ernst Bloch, *The Principle of Hope* (reprint ed. 1995), trans. by N. Plaice, S. Plaice, and P. Knight (Cambridge, MA: MIT Press, Vol. 1, 1986), p. 8.
19 Ibid.
20 Ibid., p. 7.
21 Ibid., p. 223.
22 Ruth Levitas, *The Concept of Utopia* (Witney: Peter Lang, 2011), p. 199.
23 Russel Jacoby, *The End of Utopia: Politics and Culture in an Age of Apathy* (New York: Basic Books, 1999), p. 161.
24 Sarah Wagenknecht, *Die Selbst-Gerechten: Mein Gegenprogramm – für Gemeinsinn und Zusammenhalt* (Frankfurt/New York: Campus Verlag, 2020), p. 25.
25 Thomas B. Macaulay, 'Lord Bacon, Part II, His Character and Writings', in *The Southern Literary Messenger: Devoted to Every Department of Literature and the Fine Arts*, Vol. IV (Richmond: Thos W. White, 1838), p. 78.
26 Jacoby, p. 176.
27 Rightmove PLC, 'Property Portal' <https://www.rightmove.co.uk/properties/114492785#/?channel=RES_BUY> [accessed 15 August 2022]
28 Roger Paden, 'Marx's Critique of the Utopian Socialists', *Utopian Studies*, 13(2) (2002), pp. 67–91.
29 Serge Latouche, *Farewell to Growth* (New York: Polity Press, 2009).
30 Angelos Varvarousis and Giorgos Kallis, 'Commoning Against the Crisis in Greece and Beyond', in *Another Economy is Possible: Culture and Economy in a Time of Crisis*, ed. by Manuel Castells, 1st edn (New York: Wiley, 2017), pp. 147–148.
31 Geoff Mulgan, 'The Imaginary Crisis (and How We Might Quicken Social and Public Imagination)', *UCL Demos Helsinki and Untitled* (2020), p. 23 <https://demoshelsinki.fi/julkaisut/the-imaginary-crisis-and-how-we-might-quicken-social-and-public-imagination/> [accessed 13 February 2022]
32 Elinor Ostrom, *Beyond Markets and States: Polycentric Governance of Complex Economic Systems*, Prize Lecture (8 December 2009), pp. 435–436.
33 Ostrom, p. 419.
34 Ibid., p. 411.
35 Ibid., p. 413.
36 Elinor Ostrom, *Neither Market nor State: Governance of Common-Pool Resources in the Twenty First Century*, IFPRI Lecture Series (1994), p. 1.
37 Ibid., p. 2.
38 Isle of Ulva, 'A Community Owned Island' <https://www.ulva.scot/> [accessed 27 July 2022]
39 Grounded Solutions Network, 'Community Land Trusts CLTs' <https://groundedsolutions.org/strengthening-neighborhoods/community-land-trusts> [accessed 27 July 2022]
40 Chantier de l'economie sociale (1999) <https://chantier.qc.ca/about/?lang=en> [accessed 28 July 2022]
41 Varvarousis and Kallis, p. 145.
42 Seikatsu Club (1997) <https://seikatsuclub.coop/en/about.html> [accessed 28 July 2022]
43 Pat Conaty and David Bollier, 'Toward an Open Co-operativism: A New Social Economy Based on Open Platforms, Co-operative Models and the Commons' (2014) <https://commonsstrategies.org/towards-an-open-co-operativism/> [accessed 10 August 2022]
44 Ibid.

45 Michael Bauwens, Vasilis Kostakis, and Alex Pazaitis, *Peer to Peer: The Commons Manifesto* (London: University of Westminster Press, 2019), p. 1.
46 Silke Helfrich, 'World Social Forum: 10 Years after: Elements of a New Agenda: The Commons' (2010) <https://seminario10anosdepois.wordpress.com/2010/02/17/world-social-forum-10-years-after-elements-of-a-new-agenda-the-commons/> [accessed 2 June 2022]
47 Ibid.
48 'Netarchical Capitalism', P2P Foundation, Wiki <https://wiki.p2pfoundation.net/Netarchical_Capitalism> [accessed 6 June 2022]
49 Donna J. Haraway, 'When Species Meet', *Posthumanities*, Vol. 3 (Minneapolis: University of Minnesota Press, 2008), p. 46.
50 Peter Linebaugh, *The Magna Carta Manifesto* (Berkeley: University of California Press, 2008), p. 279.
51 Anna Tsing, *The Mushroom at the End of the World* (Princeton: Princeton University Press, 2015), p. 255.
52 Ibid.
53 Ibid.
54 Ibid., p. 274.
55 Angelos Varvarousis and Giorgos Kallis, 'Commoning Against the Crisis', in *Another Economy is Possible: Culture and Economy in a Time of Crisis*, ed. by Manuel Castells, 1st edn (New York: Wiley, 2017), p. 123.
56 Ostrom (2009), p. 420.
57 Neera Singh, 'Becoming a Commoner: The Commons as Sites for Affective Socio-Nature Encounters and Co-becomings', *ephemera*, 17(4) (2017), pp. 751–776.
58 Herzler, p. 185.
59 Goodwin and Taylor, p. 181.
60 Yaacov Oved, *Two Hundred Years of American Communes* (New York: Routledge, 2017), p. 115.
61 Levitas (2011), p. 44.
62 Ibid., p. 43.
63 Levitas (2005), p. 12.
64 Hertzler, p. 225.
65 John Ruskin, *Selected Writings: The Stones of Venice II* (Oxford: Oxford University Press, 2004), p. 47.
66 William Morris, *On Art and Socialism: Essays and Lectures* (London: John Lehmann, 1947), p. 22.
67 Ibid., p. 26.
68 Mumford, p. 89.
69 Henry David Thoreau, *Walden & Civil Disobedience, and Other Writings, Authoritative Texts, Journal, Reviews and Posthumous Assessments, Criticism*, 3rd ed. (New York: W. W. Norton & Co Inc., 2008), p. 65.
70 Federation of Egalitarian Communities (2015) <https://thefec.org> [accessed 20 July 2022]
71 Adrian Shirk, 'The Romantic, Failed Experiments of American Utopias', *Yes! Journal* (2022) <https://www.yesmagazine.org/health-happiness/2022/06/01/american-communes-failed-experiment> [accessed 3 August 2022]
72 Ibid.
73 Lucy Sargisson, 'Strange Places: Estrangement, Utopianism, and Intentional Communities', *Utopian Studies*, 18(3) (2007): 393–424 <https://www.jstor.org/stable/20719884> [accessed 20 July 2022]
74 Ibid.
75 Ibid.

76 Geoph Kozeny, 'Intentional Communities: Lifestyles Based on Ideals' (1995) <https://articles-and-essays.s3.amazonaws.com/Intentional+community+/G-Kozeny-lifestyles-based-on-ideals.pdf> [accessed 20 July 2022]
77 Alexa Clay, *Like Start-Ups Most Intentional Communities Fail*, ed. by M. Benjamin (Utopia Inc., 2017) <https://aeon.co/essays/like-start-ups-most-intentional-communities-fail-why> [accessed 21 July 2022]
78 Bill Metcalf, 'Utopian Struggle: Preconceptions and Realities of Intentional Communities', in *RCC Perspectives No. 8, Realizing Utopia: Ecovillage Endeavors and Academic Approaches* (2012), pp. 21–30 (p. 25).
79 Clay (2017).
80 Ibid.
81 Sargisson, p. 393.
82 Nellie Bowles, 'Tech Entrepreneurs Revive Communal Living: Tech Entrepreneurs Create a New Kind of Shared Housing', *Sfgate* (2013) <https://www.sfgate.com/bayarea/article/Tech-entrepreneurs-revive-communal-living-4988388.php> [accessed 19 July 2022]
83 Embassy Network, 'Communities Experimenting with Culture and Commoning' <https://embassynetwork.com/> [accessed 19 July 2022]
84 Shareable, ed., *Sharing Cities: Activating the Urban Commons* (Tides Center/Shareable, 2018), p. 23.
85 Edward Relph, *Place and Placelessness* (London: Pion Limited, 1976), p. 143.
86 Lucy Sargisson and Lyman Tower Sargent, *Living in Utopia: New Zealand's Intentional Communities* (London: Routledge, 2016), p. 195.

Bibliography

Bauluz, Luis, Govind, Yajna and Novokmet, Filip, 'Global Land Inequality, World Inequality Database' (2020) <https://wid.world/news-article/global-land-inequality/> [accessed 10 January 2022]

Bauwens, Michael, Kostakis, Vasilis and Pazaitis, Alex, *Peer to Peer: The Commons Manifesto* (London: University of Westminster Press, 2019)

Bloch, Ernst, *The Principle of Hope* (reprint ed. 1995), trans. by N. Plaice, S. Plaice, and P. Knight (Cambridge, MA: MIT Press, Vol. 3:3, 1986)

Bowles, Nellie, 'Tech Entrepreneurs Revive Communal Living: Tech Entrepreneurs Create a New Kind of Shared Housing', *Sfgate* (2013) <https://www.sfgate.com/bayarea/article/Tech-entrepreneurs-revive-communal-living-4988388.php> [accessed 19 July 2022]

Burke, Monte, 'John Malone Overtakes Ted Turner as Largest Individual Landowner in the U.S.', *Forbes*, 10 March (2011) <https://www.forbes.com/sites/monteburke/2011/03/10/john-malone-overtakes-ted-turner-as-largest-individual-landowner-in-the-u-s/> [accessed 5 January 2022]

Castells, Manuel, et al. (eds.), *Another Economy is Possible: Culture and Economy in a Time of Crisis* (New York: Wiley, 2017)

Chantier de l'economie sociale (1999) <https://chantier.qc.ca/about/?lang=en> [accessed 28 July 2022]

Clay, Alexa, *Like Start-Ups Most Intentional Communities Fail*, ed. by M. Benjamin (Utopia Inc. 2017) <https://aeon.co/essays/like-start-ups-most-intentional-communities-fail-why> [accessed 21 July 2022]

Conaty, Pat and Bollier, David, 'Toward an Open Co-operativism: A New Social Economy Based on Open Platforms, Co-operative Models and the Commons'

(2014) <https://commonsstrategies.org/towards-an-open-co-operativism/> [accessed 10 August 2022]

Embassy Network, 'Communities Experimenting with Culture and Commoning' <https://embassynetwork.com/> [accessed 19 July 2022]

Federation of Egalitarian Communities (2015) <https://thefec.org> [accessed 20 July 2022]

Frank, Robert, 'John Malone Now Biggest Landowner in the U.S.', *Wall Street Journal*, 12 October (2011) <https://www.wsj.com/articles/BL-WHB-4892> [accessed 5 January 2022]

Goodwin, Barbara and Taylor, Keith, *The Politics of Utopia: A Study in Theory and Practice* (London: Hutchinson & Co, 1982)

Grounded Solutions Network, 'Community Land Trusts CLTs' <https://groundedsolutions.org/strengthening-neighborhoods/community-land-trusts> [accessed 27 July 2022]

Haraway, Donna J., *'When Species Meet', Posthumanities*, Vol. 3 (Minneapolis: University of Minnesota Press, 2008)

Helfrich, Silke, 'World Social Forum: 10 Years After: Elements of a New Agenda: The Commons' (2010) < https://seminario10anosdepois.wordpress.com/2010/02/17/world-social-forum-10-years-after-elements-of-a-new-agenda-the-commons/> [accessed 2 June 2022]

Herzler, Joyce O., *The History of Utopian Thought* (London: George Allen & Unwin, 2012, original work published 1923)

Isle of Ulva, 'A Community Owned Island' <https://www.ulva.scot/> [accessed 27 July 2022]

Jacoby, Russel, *The End of Utopia: Politics and Culture in an Age of Apathy* (New York: Basic Books, 1999)

Jameson, Fredric, *Archaeologies of the Future: The Desire Called Utopia and Other Science Fictions* (London: Verso, 2005)

Kozeny, Geoph, 'Intentional Communities: Lifestyles Based on Ideals' (1995) <https://articles-and-essays.s3.amazonaws.com/Intentional+community+/G-Kozeny-lifestyles-based-on-ideals.pdf> [accessed 20 July 2022]

Latouche, Serge, *Farewell to Growth* (New York: Polity Press, 2009)

Levitas, Ruth, 'The Imaginary Reconstitution of Society: Or Why Sociologists and Others Should Take Utopia More Seriously', Inaugural Lecture, University of Bristol (2005) <http://www.bristol.ac.uk/depts/sociology/staff/pubs/levitasinaugural> [accessed 12 February 2019]

Levitas, Ruth, *The Concept of Utopia* (Witney: Peter Lang, 2011)

Linebaugh, Peter, *The Magna Carta Manifesto* (Berkeley: University of California Press, 2008)

Lowder, Sarah K., Sánchez, Marco V. and Bertini, Raffaele, 'Farms, Family Farms, Farmland Distribution and Farm Labour: What Do We Know Today?', Agricultural Development Economics Working Paper 19-08, Rome, FAO (2019)

Macaulay, Thomas B., *Lord Bacon, Part II, His Character and Writings, in The Southern Literary Messenger: Devoted to Every Department of Literarture and the Fine Arts*, Vol. IV (Richmond: Thos W. White, 1838)

Metcalf, Bill, 'Utopian Struggle: Preconceptions and Realities of Intentional Communities', in RCC Perspectives No. 8, *Realizing Utopia: Ecovillage Endeavors and Academic Approaches* (Munich: RCC, 2012)

Morris, William, *On Art and Socialism: Essays and Lectures* (London: John Lehmann, 1947)

Mulgan, Geoff, 'The Imaginary Crisis (and How We Might Quicken Social and Public Imagination)', *UCL Demos Helsinki and Untitled* (2020), p. 23 <https://demoshelsinki.fi/julkaisut/the-imaginary-crisis-and-how-we-might-quicken-social-and-public-imagination/> [accessed 13 February 2022]

Mumford, Lewis, *The Story of Utopias* (New York: Boni and Liveright, 2008, original work published 1922)

'Netarchical Capitalism', P2P Foundation, Wiki <https://wiki.p2pfoundation.net/Netarchical_Capitalism> [accessed 6 June 2022]

Ostrom, Elinor, *Neither Market nor State: Governance of Common-Pool Resources in the Twenty First Century*, IFPRI Lecture Series (Washington, DC: IFPRI, 1994)

Ostrom, Elinor, *Beyond Markets and States: Polycentric Governance of Complex Economic Systems*, Prize lecture (8 December 2009)

Oved, Yaacov, *Two Hundred Years of American Communes* (New York: Routledge, 2017)

Paden, Roger, 'Marx's Critique of the Utopian Socialists', *Utopian Studies*, 13(2) (University Park, PA: Penn State University Press, 2002)

Relph, Edward, *Place and Placelessness* (London: Pion Limited, 1976)

Rightmove PLC, 'Property Portal' <https://www.rightmove.co.uk/properties/114492785#/?channel=RES_BUY> [accessed 15 August 2022]

Ruskin, John, *Selected Writings: The Stones of Venice II* (Oxford: Oxford University Press, 2004)

Sargisson, Lucy, 'Strange Places: Estrangement, Utopianism, and Intentional Communities', *Utopian Studies*, 18(3) (2007), pp. 393–424 <https://www.jstor.org/stable/20719884> [accessed 20 July 2022]

Sargisson, Lucy and Sargent, Lyman Tower, *Living in Utopia: New Zealand's Intentional Communities* (London: Routledge, 2016)

Seikatsu Club (1997) <https://seikatsuclub.coop/en/about.html> [accessed 28 July 2022]

Shareable, ed., *Sharing Cities: Activating the Urban Commons* (San Francisco, CA: Tides Center/Shareable, 2018)

Shirk, Adrian, 'The Romantic, Failed Experiments of American Utopias', *Yes! Journal* (2022) <https://www.yesmagazine.org/health-happiness/2022/06/01/american-communes-failed-experiment> [accessed 3 August 2022]

Singh, Neera, 'Becoming a Commoner: The Commons as Sites for Affective Socio-nature Encounters and Co-becomings', *Ephemera*, 17(4) (2017), pp. 751–776

'The Land of Scotland and the Common Good: Report' (2014), Land Reform Review Group <https://www.gov.scot/publications/land-reform-review-group-final-report-land-scotland-common-good/pages/61/> [accessed 10 January 2022]

Thoreau, Henry David, *Walden, Civil Disobedience, and Other Writings, Authoritative Texts, Journal, Reviews and Posthumous Assessments, Criticism*, 3rd edn, ed. by William Rossi (New York: W. W. Norton Co Inc, 2008)

Touraine, Alain, *Can We Live Together? Equality and Difference* (Stanford, CA: Stanford University Press, 2000)

Tsing, Anna, *The Mushroom at the End of the World* (Princeton: Princeton University Press, 2015)

Varvarousis, Angelos and Kallis, Giorgos, 'Commoning Against the Crisis in Greece and Beyond', in *Another Economy is Possible: Culture and Economy in a Time of Crisis*, ed. by Manuel Castells, 1st edn (New York: Wiley, 2017)

Wagenknecht, Sarah, *Die Selbst-Gerechten: Mein Gegenprogramm – Für Gemeinsinn und Zusammenhalt* (Frankfurt/New York: Campus Verlag, 2020)

Waggoner, Jacob, 'Land of the Free or Land of the Few? Local Concentration in US Landownership' (2020), *Harvard Mellon Urban Initiative* <https://mellonurbanism.harvard.edu/rules-engagement-forming-guidelines-design-schools-working-marginalized-communities> [accessed 5 January 2022]

Winchester, Simon, *Land* (London: William Collins, 2021)

4 The Nowherelands

From Outopia to Eutopia

Ruins as Sites for Alternative Futures

Lewis Mumford attempted in the beginning of the 20th century to establish a utopia of reconstruction which he defined as 'a vision of a reconstituted environment which is better adapted to the nature and aims of the human beings who dwell within it than the actual one'.[1]

Mumford's utopia was explicitly conceived as a eutopia (good place) that rejects the reduction of mental images to social stratifications (e.g. the 'working class') or national delimitations and instead insists on local territorial limits for communities, the use of local resources, and the occupation of spaces outside the metropolitan centers. Mumford was convinced that the spaces for the reconstituted environments would be the regions that have been left behind in the all-encompassing process of capitalist value creation. The possibilities for the creation of eutopias thus can open up when a place has lost (or diminished) its exchange value. Mumford thus predicted: 'It should not surprise us therefore if the foundations for eutopia were established in ruined countries; that is, in countries where metropolitan civilization has collapsed and where all its paper prestige is no longer accepted at paper value'.[2] In the 1920s, Mumford was convinced that the eutopian renaissance could take place in places like Germany, Austria, Russia, India, and/or Palestine. A hundred years later, Mumford's prognostics about which countries would be ruined turned out to be wrong (thus far), but his ideas about possible eutopias in 'ruined' places are a valid starting point for a reflection on ruins.

Georg Simmel looks at ruins with a more ambiguous perspective than Mumford. Instead of seeing the potential of ruined places, he underlines their retrospectivity. Simmel writes:

'Ruins however, embody the impossibility of action. They do not belong to the past world, but show the loss of that world'.[3] Svetlana Boym emphasizes the interstitial aspect of ruins:

> *Ruin* literally means 'collapse' but actually, ruins are more about remainders and reminders. A tour of ruins leads you into a labyrinth

DOI: 10.4324/9780429288180-4

> of ambivalent prepositions—'no longer' and 'not yet' 'nevertheless' and 'albeit'—that play tricks with causality.[4]

Between the remainder and reminder, between the 'no longer' and the 'not yet', ruins are thus situated in a space that forecloses the potential while at the same time revealing it. In her classic work *Pleasure of Ruins*, Rose Macaulay reflects on the different aspects of fascination that observers experience when looking at ruins:

> Since down the ages men have meditated before ruins, rhapsodized before them, mourned pleasurably over their ruination, it is interesting to speculate on the various strands in this complex enjoyment, on how much of it is admiration for the ruin as it was in its prime, [...], how much aesthetic pleasure is in its present appearance, how much is association, historical, or literary, what part is played by morbid pleasure in decay, by righteous pleasure in retribution [...], by mystical pleasure in the destruction of all things mortal [...] by egoistic satisfaction in surviving [...] by masochistic joy in a common destruction [...] and by a dozen other entwined threads of pleasurable and melancholy emotion.[5]

Since the publication of Macaulay's book, the interest in ruins has grown exponentially. In the 21st century, the preoccupation with ruins is omnipresent in sociology, art, urban geography, organization studies, philosophy, and anthropology. The pleasure of ruins has turned into contemporary 'Ruinenlust' (an 18th-century German word that described the increasing proto-romantic aesthetic interest in ruins), or in Svetlana Boym's words, 'Ruinophilia'. In its most extreme version, this 'lust' and 'philia' can even turn into 'Ruin porn'. This refers to a certain attitude of spectatorship that, as Kate Abbey-Lambertz points out, focuses and relishes 'the way decline is glamorized or exploited [...] rather than seen as part of the city's larger ills'.[6]

Whether as lust, philia, or even porn, ruins are increasingly moving into the centre space of reflection. And while the 18th-century Ruinenlust-driven audiences meditated before Roman or Greek ruins, in the 21st century the ruins that are becoming the object of contemplation and reflection are increasingly ruins of the recent past, ruins of contemporary societies. Caitlin DeSilvey and Tim Edensor diagnose: 'We seem to be in the midst of a contemporary Ruinenlust, which carries strange echoes of earlier obsessions with ruination and decay'.[7] Obsession about ruins has turned factories, former communist buildings, urban wastelands, derelict train networks, and abandoned military structures into sites of interest for the contemporary ruinophile. The broadened gaze on ruins has also inspired an ever-growing vocabulary on the phenomena of contemporary ruins. As DeSilvey and Edensor observe:

> Work in this area has generated a bewildering array of neologisms and conceptual tags: 'unofficial countryside' (Mabey, 1974), 'edge-lands' (Farley and Roberts, 2011; Shoard, 2000), 'places on the margin' (Shields, 1992), 'terrain vague' (De Sola Morales, 1995), 'dead zones' (Doron, 2000), 'anxious landscapes' (Picon, 2000), 'parafunctional space' (Papastergiadis, 2002), 'voids' (Cupers and Miessen, 2002), 'landscapes of contempt' (Girot, 2005), 'indeterminate spaces' (Groth and Corijn, 2005), 'awkward spaces' (Jones, 2007), 'found spaces' (Rivlin, 2007), 'ambivalent landscapes' (Jorgensen and Tylecote, 2007), 'drosscapes' (Berger, 2006), 'loose spaces' (Franck and Stevens, 2007) and 'urban interstices' (Tonnelat, 2008).[8]

Apart from the focus on contemporary ruins, what differentiates the 21st-century obsession with ruins from the proto-romantic, romantic, and post-romantic meditation on ruins in the 18th and 19th century?

In 1796, when the French painter Hubert Robert, who was nicknamed Ruin Robert (Robert des Ruines) by his contemporaries, revealed his painting 'Imaginary View of the Grande Galerie of the Louvre in Ruins' to the audience, they could interpret his vision of destruction as an aesthetic play. In the capriccio genre of paintings painters, like Piranesi, experimented with imagined scenarios of destruction. So, while looking at the ruins of the Grande Galerie of the Louvre, nobody could ever realistically assume that such a scenario would ever come into being, the painting was thus a mere aesthetic fantasy. John Soane and Joseph Michael Gandy produced a similar vision of hypothetic destruction in 1830, in their 'A Bird's eye view of the Bank of England', the mighty building complex was shown in a state of devastation. In the drawing, more than 90% of the Bank of England was destroyed, the roof structures collapsed, the walls broken down; everything turned into rubble and debris. Again, like in Robert's case, the destruction is a fantasy that serves as an assurance of the stability of these iconic structures, and possibly, to emphasize their future historic value by portraying these buildings in the same way as classic Greek and Roman edifices.

The obsession with the Greek and Roman ruins dominated a lot of the aesthetic and philosophic contemplation in the 19th century and also influenced architecture in the 20th century. Albert Speer developed the Ruinenwerttheorie, and when he became the architect of the Third Reich, he lobbied the idea that the key buildings of the Nazi regime should be constructed in such a way that they would look impressive even hundreds of years after their destruction. The idea was that really significant architecture should leave a lasting legacy, like the Roman and Greek ruins.

In the aftermath of the global destruction in the Second World War, under the impression of the atomic bomb explosions in Hiroshima and Nagasaki, the Japanese architect Arata Isozaki symbolically buried the idea of lasting legacy. He wrote: 'Future cities are themselves ruins. Our contemporary cities … are destined to live only a fleeting moment. Give up their energy and return to inert material. All of our proposals will be buried'.[9] The monumental ideas of the playful destructions of Robert and Soane, which ultimately were manifestations of the ambition of a lasting legacy and the megalomania of Albert Speer's Ruinenwerttheorie, were thus invalidated by the sheer potential of destruction that manifested itself in the Second World War.

Henceforth, while the Antic ruins still maintained their value as sites of touristic attractions, the ruin gaze shifted. In 1967, Robert Smithson made his 'dérive style' excursion to Passaic with an Instamatic 400 camera. He discovered in the New Jerseyan suburban landscape nondescript bridges, pipes in the water, or sandboxes, photographed them and thus turned them into 'monuments of Passaic'. The zero panorama he discovered, what he called 'ruins in reverse', buildings that are falling into ruins before they are built because they have no relationship to time, no past, and no meaning. Without reference to time and history, everything turns grey. Or, as Smithson describes it through the example of the nondescript sandbox:

> Picture in your mind's eye the sandbox divided in half with black sand on the one side and white sand on the other. We take a child and have him run hundreds of times clockwise in the box until the sand gets mixed and begins to turn grey; after that we have him run anti-clockwise, but the result will not be a restoration of the original division but a greater degree of greyness and an increase of entropy.[10]

Smithson's contemplations of the nondescript ruins in reverse and the increase of entropy were still written and photographed without the awareness that even these artefacts of urban growth and decay would soon spread in a more spectacular way not only in the United States but all over the world. The riots and protests at the end of the 1960s and the 1973 oil crisis were indicators that the growth model of the post 1945 Western economies would not result in an infinite trajectory of economic and social progress. Economic devastation and the disintegration of cities were all of a sudden also conceivable not only in faraway African or Latin American countries but also in the very heartland of Western capitalism: the United States. In the process of deindustrialization and suburbanization, some of the major American cities started to disintegrate, shrink, fall into social despair, and thus, first imperceptibly, then strikingly obvious, (at least partially) turned into urban ruins. The breakdown of the Soviet Union did not only produce wastelands on the territory of the post-communist countries, but also added another catalyzer for the destruction of some American cities. The post 1991 globalized world order facilitated the dismantling of the leftover industries and thus turned many of the former economic powerhouses of the so-called 'Rust Belt' into disaster zones. The most iconic example, the poster child for the new ruins, was the city that was, until then, commonly seen as one of the biggest economic success stories in the post-World War II era: Detroit.

If in the 21st century, Silicon Valley is metonymic for the success story of American capitalism, in the second half of the 20th century Detroit held that position. The Google, Apple, and Facebook of that time, often qualified as the 'old' economy were General Motors, Chrysler, and Ford. The decline of Detroit is connected to different waves of deindustrialization that, as Kaeleigh Herstad points out, started already in the period between 1948 and 1967[11] when, due to automation, 130,000 jobs in the car industry were eliminated. The decline was a slow and gradual process, which started already in the rose-tinted times of the American economic boom of the postwar era. The process of deindustrialization accelerated in the 1980s and 1990s, and by the turn of the millennium, it became obvious that the boom turned into gloom. Chrysler's and GM's bankruptcy in 2009 was followed by Detroit's bankruptcy in 2013. The point of no return for Detroit was already reached in 2011, when the US Census Bureau stated that the city's population had fallen to 713,777 inhabitants, while Detroit's financing was based on a minimum tax base of 750,000 people. According to the 2020 census, Detroit has 639,111 residents. During the post-war boom, 1.85 million people lived in the city. Formerly the fourth largest city in the United States, Detroit fell to position 23 in a recent American city ranking.[12] The traumatic decline also radically transformed the urban landscape. For the first time, the ruins were not displaced in time and space but they were looming in the present of the heartland of capitalism. A city shrinking down to one-third of its population, real estate that becomes unreal estate because it loses all its (exchange) value, and an urban infrastructure that, due to the crisis of public finances, is left in complete decay. Sometimes the details of reports that are claiming the recovery of a city are most shocking, e.g. Detroit Future city states that in '2016, the city experienced 3400 building fires, 300 fewer than 2015'.[13] So, every year in Detroit more than 3000 houses are burning. In addition, there are 78,506 decaying houses. Despite having a partially uninhabitable housing infrastructure, the report boasts that 'that 30% of Detroit's housing, or 109,788 units, are vacant'.[14] But the developments in Detroit are not symbols for an empire in decline, such as the Roman and Greek ruins. The urban decay of the former industrial hub is rather a signifier of a certain style of capitalism in decline. The Western industrial model, in its Detroit version or in the version of the German automobile hub Wolfsburg (the headquarters of Volkswagen), was based on a high demand for a workforce, relatively high wages for a large number of workers, trade unions, and a commitment of the company to a specific location (and thereby high amounts of tax contributions). The Silicon Valley model has replaced this with a digitalized value creation that is increasingly less dependent on workforce, independent of location, and non-committal on tax payments. Thomas Sugrue underlines that

> Detroit's postwar urban crisis emerged as the consequence of two of the most important, interrelated, and unresolved problems in American history: that capitalism generates economic inequality and

> that African Americans have disproportionately borne the impact of that inequality.[15]

Detroit, 80% of its population being African American, is showing the disproportionate impact of economic crisis on the non-Caucasian population. But the impact is gradually tangible for all American citizens who are not part of what Occupy Wall Street called 'the 1%'.

With more than 70,000 decaying houses and more than 3000 burned down houses every year, Detroit became the mecca for ruin tourists. The Robert Smithsons of the 21st century were not flaneurs on the dérive but voyeurs of decay. Photographers like Andrew Moore, Yves Marchand, and Romain Meffre produced coffee-table books in which the ruins of Detroit were aestheticized and mystified in order to produce shivers of excitement for upper middle-class audiences in London lofts, New York townhouses, or suburban McMansions. Most notably absent from many of the glossed over photos were people. Detroit ruins were disembodied from the social problems of the city. The 639,111 residents were deleted from the city picture. As John Patrick Leary argues:

> So much ruin photography and ruin film aestheticize poverty without inquiring of its origins, dramatizes spaces but never seeks out the people that inhabit and transform them, and romanticizes isolated acts of resistance without acknowledging the massive political and social forces aligned against the real transformation, and not just stubborn survival, of the city.[16]

Leary emphasizes that photos cannot adequately explain the origins for Detroit's decline, they rather represent 'the spectacular wreckage left behind in the present'.[17] The decades of deindustrialization, housing discrimination, suburbanization, drug violence, municipal corruption, and misled urban renewal projects are thus never part of the picture.

Partially attracted by the 'ruin porn', the vacancy, and the low rental prices, Detroit became not only the mecca for photographers on the transit but also for artists on the lookout for affordable living spaces. The loss of exchange value of many Detroit factory ruins or detached houses created a new form of use value. A lot of artists and art collectives started to move to the former 'Motor City', for example, the former New York-based Galapagos Art Space. In 2013, the group proudly announced on their website that they bought nine buildings with the total size of more than 180,000 square meters for the price of a small apartment in New York City. Robert Helmes, the Executive Director of Galapagos, explains the reasons for the move: 'actually being an artist in New York City costs you a full time-career in another industry, then the best and brightest – the ones our meritocracy would obviously miss the most – won't allow their work to suffer just to be among our tall buildings'.[18] The drive towards the 'ruined' Detroit is thus explained by the hope for a space of possibility. It is not without irony that Galapagos' exchange of a (dreary) apartment in New York for a 180,000 square-meter building in Detroit is animated by the hope to find the freedom and free space to be artistically creative. While certainly nobody would call Detroit 'a principality of utopia' the Galapagos certainly left the 'acre' in New York behind. Unfortunately, the story of the Galapagos Art Space did not end well. On 9 January 2019 on the website of the Detroit Free Press, J.C. Reindl reported that Galapagos sold the Detroit property to the real estate firm Bedrock. The spokesman of the Bedrock company praised the great potential of the property company and concluded: 'We look forward to determining our development vision for the property'.[19] According to the article the reason for the sale was largely financial mismanagement from the side of Galapagos. But, of course, this can also serve as a cautionary tale that in real-life adventure stories, a happy ending is never guaranteed. Certainly, the sale of the 180,000 square meter building to a real estate company is a continuation of the focus on the same form of property accumulation that created the reason for the transfer of Galapagos from New York to Detroit in the first place.

The coffee-table books with the cool commodifiable gaze on ruins and the property accumulation drive of Galapagos Art Space are just some examples of gentrification in the middle of decline. As Bellhop, a company for moving services, points out on their website, Detroit has '5 up and coming neighborhoods'.[20] The company lists Fitzgerald ('the community regularly gets together for bonding events like mural paintings and picnics'), North End ('quality housing stock, great neighborhood amenities, and affordable price points (…)Investors don't expect the low cost of these houses to last, though'), Southwest ('affordable houses, an expanding shopping and dining scene, and the promise of growth'), Corktown ('This charming neighborhood is lined with spacious sidewalks and character-rich, wood-framed houses'). The top of the crop of this top 5 list is West Village, Bellhop's website explains:

> Prices have blown up in the trendy West Village corner of the city. But new buyers keep piling in, making this neighborhood still a great option for investment. West Village is becoming less and less affordable for a lot of locals, but prices are rising for good reason. […] Packed with millennials, West Village offers its residents easy access to downtown and the River Town Warehouse District. The neighborhood is walkable, families feel safe and at home here, and investors have their eyes on all the action. […] Nearby, luxury condos are going up.[21]

Kaeleigh Herstad describes the process of reclamation of Detroit heritage as a commercialization of salvaged items. Blighted neighbourhoods, ruins, and abandoned spaces are salvaged and used for products to be sold to the new Detroiters, the middle-class target group to whom Bellhop is showcasing the 'up and coming neighborhoods'. Companies like Home Eyewear are making sunglasses, using reclaimed wood from ruins, stamping them with the street name where the wood was found, and selling the recycled product for $168. Workshop, another Detroit company, uses reclaimed wood from ruins and builds furniture with a price range from $255 to $2250. The local company Leadhead Glass recycles salvaged glass and woods to build handcrafted terrariums, ranging from $80 to $230. Herstad emphasizes that there is a pronounced divide between residents of the affluent neighbourhoods (Downtown/Midtown/Suburbs) and the rest of the city and indicates:

> When the per capita income in Detroit is less than $15.000 a year and more than 40 percent of residents live under the federal poverty line, spending $168 for a pair of sunglasses or $1500 for a table made of reclaimed wood is unlikely to be an option for most residents.[22]

Of course, the target group for these products are the affluent residents of West Village and Corktown who purchase these reclaimed and handmade objects, possibly because they represent a piece of Detroit authenticity or because they are signs that the city is developing.

The affluent downtown neighbourhoods are just one part of the 'Detroit comeback' story.

The symbol of the revitalization of Detroit is a building that, a few years ago, was called 'the Eiffel tower of ruin photography'[23] (and was thus prominently featured in the coffee-table books by Moore, Marchand, and Meffre): Michigan Central Station.

When ruin photographers are looking for a good motive, they can safely scratch Michigan Central Station from their list. In a twist of irony, the Ford company, one of the former Big Three companies in Detroit's Motor City Boom, in collaboration with Google, has started to renovate the station building. The hundreds of broken windows and the dilapidated arrival hall, which were featured so prominently in many of the ruin photos, have been renovated, the building face restored, and in 2023 'Michigan Central' ('station' has vanished from the name) will open up as an innovation hub. As the website of this global mobility platform proudly claims: 'Michigan Central will include new and revitalized buildings, a first-of-its-kind mobility testing platform, multiple outdoor plazas, open spaces and 1.2 million square feet of commercial public space'.[24]

Michigan Central thus is telling a revitalization story, a deindustrialized city coming back from the ruins. Jefferson Cowie and Joseph Heathcott contest such a seamless transition when they write:

> The point of departure for any discussion of deindustrialization must be respect for the despair and betrayal felt by workers as their mines, factories, and mills were padlocked, abandoned, turned into artsy shopping spaces, or even dynamited. While economists and business leaders often

> speak in neutral, even hopeful, terms such as 'restructuring' 'downsizing' or 'creative destruction' metaphors of defeat and subjugation are more appropriate for the workers who banked on good-paying industrial jobs for the livelihoods of their families and their communities.[25]

Cowie and Heathcott underline that the 'very set of political rules that created the industrial order that we once took to be permanent provided the means by which corporations could dismantle that order'.[26] In relation to the new Michigan Central, this means that Ford which was the pioneer of the American car industry and, at its peak, created 90,000 jobs in the Detroit region's River Rouge plant alone, is now announcing the creation of 5000 jobs in the complex of the former railway station. In a shrinking city that has lost two-thirds of its population, Ford is thus bringing back one-eighteenth of the jobs it used to provide in only one of its many car factories in and around Detroit. The new 'miniature' Detroit is thus, in its gentrified version, again designed as a poster child, this time not for ruination but for regeneration. But the shiny buzz is only for the selected few (millennials). Forty percent of Detroiters and 57% of the children in Detroit are still living below the poverty line.[27] To the poor, the motto of the city of Detroit, 'Speramus Meliora; Resurget Cineribus' (*We hope for better days; it shall rise from the ashes*), sounds like a hollow promise.

When Camilo José Vergara, a photographer who started to document the ruination of American cities long before it became fashionable, looked at the Michigan Central Station in 1999, he realized that many locals also stopped to contemplate the massive building. Vergara observed that in Detroit 'nothing arrives or leaves from the former station; only the imagination travels'.[28] At the time, the ruins of Detroit contained the seed of a possibility. Vincent Carducci dreamed of an art of the commons, he observed 'a tendency within contemporary Detroit art that has emerged in those spaces where the distinctions between public and private seems to have dissipated as part of the process of demassification of the city's core, which has taken place over the last four decades'.[29] The possibility of a dissipation of the distinctions of public and private, the train of imagination taking its course in a leftover space of capitalism, the space of possibility to create a new city – all of these ideas dominated the world of ideas of some Detroiters in the early 2000s. Initiatives that reclaim the abundant abandoned city space were developing and had a positive influence on neighbourhoods and communities. For example, the Michigan Urban Farming Initiative has helped to develop 1400 community farms in Detroit that produce enough fruit and vegetables to supply 20% of the city residents with food. Neighbourhoods transforming into agrihoods were encouraging signs of a positive transformation.

But in the absence of public financing, soon the private investors came to occupy the space. With investors on board, business incentives in the backhand and money generation as a primary goal, the train of imagination soon had no more station to depart from. Almost 20 years later Vergara, upon observing the ongoing gentrification tendencies in Detroit, came to conclude that 'Michigan Central Station had never looked better and it will never look better than when it was at its most neglected'.[30]

In Detroit, gentrification that goes hand in hand with inequality and high levels of poverty is accompanied by demolition. For Jason Hackworth, the blight removal is motivated by the desire to visually bury the social problems instead of solving them. Hackworth writes: 'Demolition, within this paradigm, is not about community building as much as it is about making the problem disappear. Already-disinvested neighborhoods are framed as already moribund or dead. Accelerated demolition is merely a form of state-sponsored burial'.[31] So while poverty does not disappear the new Detroit, in its 'destroyed and saved by capitalism' version, its traces are buried and thus made invisible.

The new shrunken Detroit is in a paradoxical situation: it still needs to project the growth ideal of capitalism despite the fact that the whole city was downgraded and diminished. As Leary explains, the 'shrinking city [...] defies much of what we think we know about cities and their development, at least as long as 'growth' and 'development' are the only measures of vitality'.[32] In Detroit, the policies of redevelopment of recent years have reasserted the same capitalist strategies of growth production that have led to the ruin. The imaginary that has led to decay is now, in its version of downsized expectations (drastic reduction of the possibility of earning a livelihood for the majority of the population, privilege for the few) managing the aftermath of the decay.

Anna Tsing has a different vision for the aftermath of economic breakdown. Tsing observes:

> This is a story we need to know. Industrial transformation turned out to be a bubble of promise followed by lost livelihoods and damaged landscapes. (...) If we end the story with decay, we abandon all hope – or turn our attention to other sites of promise and ruin, promise and ruin.[33]

The promise and ruin narrative of capitalism is getting another chapter in Detroit, which shows a strange level of resilience to the lessons the city should have learned from its ruination trajectory. Tsing sees more potential in the ruins, which the capitalist value generation process produces in growing numbers, and underlines that they 'can be lively despite announcements of their death; abandoned asset fields sometimes yield new multispecies and multicultural life'.[34] As she observes, in 'a global state of precarity, we don't have choices other than looking for life in this ruin'.[35] These ruins can become places of salvage, of taking advantage of value produced without capitalist control. So, salvage in this case does not mean the commodification of salvaged goods by Detroit companies like Home Eyewear or Leadhead Glass but the generation of use value and encounter value. In

Detroit, spaces of (potential) abundance are yet again turned into spaces of scarcity.

The potential of ruins is also emphasized by Caitlin DeSilvey and Tim Edensor who see them as excessive spaces with a wide scope for direct engagement and imaginative historical reconstruction. DeSilvey and Edensor explain: 'Derelict spaces are proffered as sites that, by contrast, are replete with latent possibilities and greater scope for conviviality, experimentation and expression'.[36]

If we see the post-bankruptcy Detroit as an experiment, then it is clear that it was dominated by the neoliberal imaginary, in which private investments lead to segmented spaces, borders between the privileged and the poor, and exclusion. Common spaces are reprivatized and transformed into value-generation projects for the wealthy. This amounts to a 21st-century remix of land enclosures of the 16th century. The irony here is that the commons in Detroit were the ruins of capitalism, the landscape opened up but the mind scape is still locked in the enclosure; and, thus, the enclosure comes back. The lack of imaginary turned the space of possibility into a space of re-enclosure. Instead of experimenting with new forms of cohabitation, community-building, and value creation, the neoliberal experiment with human livelihood simply continues in an unmodified way.

The battle of the story for a more commons-oriented solution to the revitalization of shrinking cities was thus lost. Potentially the symbolism of Detroit, one of the former hotspots of capitalism, turning into the first post-capitalist space was too damaging for American corporations. To see the city where Henry Ford test drove his first car in 1896 turn into an experimental playground for post-capitalist ways of living was probably too horrifying for the Ford Corporation. So, the involvement in Michigan Central was certainly not prompted by sentimentalism but by image control. The reinvestment process in Detroit was certainly motivated by the prospect for profit but also in order to stop the train of imagination. Detroit had too high a symbolic value to be left to experimentation.

So, as opposed to telling a possible 'ruination and reclamation by citizens' story – so a hero journey from a grassroots level, an empowerment narrative – the story that won in the battle around Detroit is a storyline centred around the archetype 'from riches to rags, from rags to riches'. The hero of the story: American business. And the moral: capitalism can recover. The famous dictum by former mayor Coleman Young 'Detroit today is your town tomorrow' has become ambiguous: what he referred to was the ruination of the city, but does this also include the socially discriminatory recovery as a downsized Detroit?

In Detroit, Mumford's idea that the foundations for eutopia can be established in ruined places was thus pushed aside in favour of 'more of the same'. But the potential of ruined places, of abandoned places with abundances of space, of potential sites for salvage and reclamation is visible all over the world. The questions of how to create commons, potentially in the form of community land ownership, how to establish new forms of co-operative living, how to generate new forms of exchange are still open. Ruined spaces, or more properly defined as leftover spaces, since a lot of these places are potentially inhabitable, offer infrastructures that might be broken but salvageable. Whereas many intentional communities had to build their living environment from the scratch (and thus invest high amounts of work and money), the leftover spaces, the ruins of capitalism, are spaces of reconstruction. The fact that these spaces used to be (fully) inhabited are already an indicator that they are (potentially) liveable. Recycling spaces is also more resource-friendly than building from the scratch.

As the Detroit case has shown, if the interest of private investors is still existent, and the public sector has no resources to support the grassroots movements in their growth, the potential for reconstruction, reclamation, and salvage is very low. In Europe, due to the higher involvement of public institutions, the prospect of collaboration through polycentric governance, the creation of common pool resources, and self-organized local institutions, is much greater.

In order to find these spaces of possibilities, it is time to take a look at the European map with our focus away from the centers of demographic growth.

The Topoi for Reconstruction

Karl Marx has never been bigger. His head is 7.1 m high and is enthroned on a pedestal with a total height of 13 m. Marx's head, 40 tons heavy is the second biggest bust in the world (only surpassed by Lenin's head in the Siberian town of Ulan-Ude). He scans the grey square of a town that used to carry his name. Old people in beige and brown-coloured jackets slowly trudge forward, some with the help of a rollator, others carry the load of Lidl plastic bags, along the Brückenstrasse. This used to be Marx's town, it was named after him, but then the winds of change came. Today the cold autumn wind blows flyers that advertise a solarium over the street. Welcome to Chemnitz, the third biggest city of the Saxony region in Germany.

While Karl Marx's bust is still the same size since its inauguration in 1971, the city around it has dramatically shrunken. A few months after the fall of the Berlin Wall, its residents voted with a big majority to leave the name 'Karl Marx Stadt' to the rubble of history and return to the historic name of Chemnitz. And then in the following years, many left the city. Chemnitz was inhabited by 307,543 people in 1991; in a 2021 census, the city counted only 243,646 remaining residents.[37] This is not a decline of Detroit dimensions, but the demography of the remaining residents is making a further decrease in the future years biologically inevitable: Chemnitz has only 22,087 residents in the age group from 0 to 10 years old, but 85,443 inhabitants are over 60, and many of them are over 80.[38]

When the town becomes the European Capital of Culture in 2025, it can indeed project a future idea of Europe to its visitors. Chemnitz as showcase for an ageing continent. The former 'Saxon Manchester', an industrial powerhouse of the 20th century, the pride of the GDR government who honored the city with the name of the most significant German communist, Otto Grotewohl, the GDR prime minister of his time, praised its inhabitants in 1953: 'The people who live here do not look back, but look forward to a better future'.[39]

Today the square in front of the Karl Marx bust is as grey as the hair of the majority of the Chemnitz population. But not only Chemnitz is aging. In the European Union the median age is 43. Europe is the continent with the oldest inhabitants in the world. The average European is 12 years older than the average citizen of the rest of the world.[40] And the further tendency towards demographic decline is inevitable. A fertility rate of 2 (children per family) is needed for mere replacement of the population, but in 2020, the rate in the European Union countries was 1.5.[41]

'Europe's demographic time bomb' has garnered growing attention in the mainstream media, the Financial Times, for instance, based on UN population projections, came to the conclusion that by '2035, there will be about 50 m fewer people of working age in Europe than in 2010'.[42]

The consequences of this combination of high median age and low fertility rates will have a profound impact on the economic, social, cultural, and political life of the European continent. As Dirk J. Van de Kaa observes:

> The fundamental changes in fertility and family formation in industrialised societies after the mid-1960s were, as Lesthaeghe and I sketched, truly revolutionary and occurred with surprising suddenness and simultaneity. It is the unprecedented low level of fertility, coupled with the increased expectation of life at advanced ages, that exacerbates the ageing process and makes migration the obvious variable to provide compensation.[43]

Van de Kaa thus sees a second demographic transition, the first transition being the 'révolution démographique' analyzed by Aldophe Landry in the beginning of the 20th century. This first demographic transition was characterized by a combination of a high mortality level combined with a high level of fertility and thus to a limited but sustained reproduction of the population. The second demographic transition, Van de Kaa claims, is characterized by the combination of aging populations and low fertility rates, what he describes as the revolutionary and surprising trend after the mid-1960s. Van de Kaa explains: 'The fundamental principle of the first transition was, in Landry's terms 'la rationalisation de la vie'; the world was demystified. The fundamental principle of the second is, in my view, the right to self-realisation granted to each individual; it is the demystification of social control'.[44]

In Europe, the corrosive combination of aging populations and below reproduction fertility rates leads to a shrinking continent. But as the example of Detroit illustrates, the shrinking process is unequally distributed. The demographic centres, Paris, London, Madrid, and Stockholm attract the young upwardly mobile and thus project dynamism and demographic growth, whereas the periphery is gradually deserted.

Regions which are classified as less developed (NUTS 2) regions by the European Union have a GDP of less than 75% of the EU average. The NUTS 2 'atlas' reveals that most of these regions experience the self-reinforcing process of economic decline and demographic decline.

Some regions are thus locked in a vicious circle. For instance, the Bulgarian region of Severozapaden which consists of the municipalities of Vidin, Montana, Lovech, Vratsa, and Pleven, is in a deadlock of high unemployment, economic stagnation, and subsequent emigration. According to Eurostat, Severozapaden has the highest Not in Employment, Education or Training (NEET) rate in the EU[45]. NEET basically refers to young people without any chance or prospects in the labour market. In the North Western Bulgarian region, 46.5% of 18–24-year-olds are condemned to a NEET existence. Another EU report came to the sobering conclusion that Severozapaden has an 'unfavourable labor force profile', a 'deteriorated demographic', and is 'the least attractive to foreign investment'.[46] Hence the region has the highest unemployment rate in Bulgaria. With a total population of 728,127, Severozapaden has lost 19.2 per 1000 inhabitants in just one year.[47] Life expectancy in the region, at 73.6 years, is among the lowest levels in Europe (the EU average is 83.8 for women and 78.3 for men). Living in Severozapaden thus means living at least half a decade less than the average European. Possibly the region could become an example for the third demographic transition – low birth rates, low life expectancy. In times of crisis, Severozapaden's precarious situation is even more aggravated: during the Covid wave of the 19 October to 27 December 2020, Severozapaden had the highest death rates on the continent with 701.9/100,000 dying due to the virus. Once written off as economically unattractive, the vicious downward spiral just drags whole areas into gloomy obsolescence. The biggest city of the region, Pleven, has a large karst valley nature park called Kaylaka in which the Roman ruins of Storgosia remind one of the declines of past civilizations. If the downward spiral of Severozapaden is not stopped, the place could become a new hunting ground for the contemporary ruinophiles.

With Severen Tsentralen, Bulgaria has another crisis-ridden region in its territory. In Serbia, the Vojvodine Region in the north of the country shows similar tendencies of interlocked economic and demographic decline. Western Greece, Dytiki Ellada, is on the path of structural decline in the same way as the Western Latvian region of Kurzemes (Courland). The biggest city of Courland is Liepaja.

Located at the Baltic Sea coast, with long beaches and lush forests, Liepaja was a popular sea bath at the end of the 19th century and became a centre for heavy industry in metal construction and oil extraction in the 20th century. Like many former Soviet industrial cities, Liepaja experienced the post-communist era as a decline orchestrated by closing industries, lack of new job opportunities, emigrating high school graduates, and demographic decline. The town was thus shrinking, from 82,241 people in 2006 to 68,948 people in 2018.[48] Despite its undeniable potential to return to the sea bath tradition and tourist destination, and despite its ice-free harbor with a ferry connection to Germany and Sweden, Liepaja has not managed to develop a sustainable economic model in the post-industrial era. Liepaja, like Latvia, is experiencing a demographic decline that is a combination of low birth rates, high mortality, and a significant number of young Latvians who are leaving the country. As a result, in 2021, Latvia had the lowest birth rate recorded over the past hundred years, and, for the first time, the number of deceased was twice as high as the number of babies.[49]

Italy equally suffers from a low birth rate. In addition, the regional imbalances are much bigger than, for instance, Latvia. The traditional divide between the prospering North and less developed South has become stronger. Regions like Abruzzo, Puglia, Molise, Sardinia, and Sicily experience strong decreases in populations, due to low birth rates, high emigration, and high mortality rates. In these regions, the age structure increasingly changes in disfavour of the young.

As the NUTS map indicates, the European Union has a growing number of towns and villages which are leftovers, shrinking cities, places of abandonment. Despite the fact that the EU has a structural fund for these regions, money is not improving the life in Liepaja, Castiglione di Sicilia, Pleven, Novi Sad, or Nafpaktos. What could give life to these towns again are people who are interested in developing projects there, who can find spaces of possibility that the metropolitan centers cannot provide (anymore), and who find good reason to discover abundance in abandon.

In 2021, EU-wide, 13.1% of the 15–29-year-olds are NEETs,[50] thus not employed or in education or training. One-sixth of the young generation is at risk of becoming the collateral damage of a misadjusted economic and educational policy. From an economic point of view, many of these young people are at risk of becoming superfluous. With the rise in automation, the increasing replacement of human beings by AI or robots, the number of 'superfluous' people is likely to rise in the coming years. In order to prevent the marginalization spreading from the periphery to the centre, and in order to provide livelihood to people and reactivate places on the edge of abandon, it is important to recall what the most important assets in an economy should be. Joseph Stiglitz pointed out that, after all, business is not an end unto itself, excessive inequality puts the macro-economic order at peril, and thus 'people are a country's most important assets, and it is important to make sure that every person can perform up to their potential'.[51] One of the most important questions of the next decades will be whether it is possible to embrace the technological and demographic change without further marginalizing increasing sections of society.

Terms like 'technological unemployment', 'second machine age', and 'global useless class' are the new projections of either fear or hope that have creeped into media, politics, business, and academia. In Germany, the term Industrie 4.0 (or Fourth Industrial Revolution – 4IR) is the synonym for a new era of economic growth which also has the potential to alleviate the social and demographic problems that the oldest continent on the planet must confront in the 21st century. All the problems of inequality, unemployment, aging populations, shrinking towns, and environmental degradation are thus set for a technological fix. The optimists are convinced that the new era of automation will generate so much wealth that the prognosis that John Maynard Keynes made in the aftermath of the Second World War will finally become reality. Keynes wrote: 'For the first time since his creation, man will be faced with his real, his permanent problem - how to use his freedom from pressing economic cares ... to live wisely and agreeably well'.[52] The only problem: in the same way that Keynes could not predict the disruption of the Western industry-based welfare system, many of the techno-optimists fail to acknowledge the potentially devastating effect of the new level of automation on the labour market and on society.

The projections on what the new world of automation will bring are thus contradictory. Comparative studies on the impact of technological change on unemployment are pointing at the fact that, historically, obsolete jobs have been replaced with new jobs and thus not disrupted the labour market in too radical ways. As Hötte, Sommers, and Theodorakopolous point out:

> the labour replacing effect of technology is typically offset by a range of compensating mechanisms that lead to labour creation or reinstatement. Hence, there does not appear to be an empirical foundation for the fear of technology-driven massive unemployment.[53]

But the authors also recognize that past empirical studies on the impact of widely diffused technologies cannot project the impact of technologies that are going beyond mere automation of manual tasks. Hötte et al. underline that 'empirical evidence on the impact of artificial intelligence, quantum computing, virtual reality, biotechnology, nanotechnology, renewable energy, and other emerging technologies that will soon impact our economy remains limited'.[54]

Thus, the basic message of some studies would still be what the National Commission on Technology, Automation, and Economic Progress stated in the 1960s when the then president Lyndon Johnson inquired about the impact of automation on the labour market. The conclusion of the commission was 'the basic fact that technology destroys jobs, but not work'.[55]

For the next decades that would simply mean, as Cognizant suggests, that there will be new jobs in the future that will replace the old ones. For example, data trash engineers, 'who can identify unused data [...], clean that data and feed it into machine-learning algorithms to find hidden insights by not only increasing how much data is collected, but also improving the data quality'.[56] Or juvenile cybercrime rehabilitation counselors, who 'will work closely with young cyber offenders to rehabilitate and introduce them to ethical cyber use'.[57] Or they can work as Joy Adjutants, whose 'humaneness, counseling and intervention will work in concert with the JOOYS platform to help customers break the inertia of where to begin by sequentially sorting clothes, papers, miscellany and – finally – sentimental items'.[58]

As a report by the McKinsey Global Institute indicated 'about half of all work activities globally have the technical potential to be automated by adapting currently demonstrated technologies, the proportion of work actually displaced by 2030 will likely be lower, because of technical, economic and social factors that affect adoption'.[59] The report underlines the significant benefits for users, businesses, and economies and the potential for productivity growth and economic growth. However, the authors also underline the scale of transformation which could exceed the scale of historical shifts, like in the 18th and 19th centuries (from agriculture to industrialization). Automation can thus lead to further income polarization and a decline of middle-wage occupations. In addition, the report suggests that 'by 2030, 75 million to 375 million workers (3–14 percent of the global workforce) will need to switch occupational categories'.[60] In addition, 45 million Americans, one-quarter of the country's total workforce could lose their job by 2030.

The 2020s will thus bring a seismic shift in the labour market, and the open question is how much unemployment the automation will produce. Some projects like the Alibaba Flying Zoo Hotel in Hangzhou are showcasing a future without humans. The hotel is largely robot-driven, the reception is made by a robot using a face biometrics system, the room service of the 'Alibaba Future Hotel' is operated by service bots, and at the bar a robot is mixing the cocktails. It is questionable if the workforce might be replaced in the hospitality sector. If the Flying Zoo Hotel should turn out to be a successful formula, maybe they can train to become joy adjutants or juvenile cybercrime rehabilitation counselors.

The World Economic Forum also warned that 'a new generation of smart machines, fueled by rapid advances in artificial intelligence and robotics, could potentially replace a large proportion of existing human jobs'.[61] According to the authors of the study, up to 30% of existing jobs across the OECD could be at potential risk of automation.

As Elon Musk recognized in an interview with CNBC in 2016, 'computers, intelligent machines and robots seem like the workforce of the future. And as more and more jobs are replaced by technology ultimately will be sustained by payments from the government'.[62] The call for government intervention by a prolific entrepreneur like Musk is a surprising twist to the debate. This could be an indicator that possibly the disruptive impact of the automation processes is much bigger than projected. After all, government interaction is generally seen as a remedy in times of profound crisis (e.g. the economic crisis of 2008 and the Covid pandemic of the early 2020s).

Yuval Noah Harari, paints a more pessimistic picture of the impact of automation. He warns of the extreme rise of inequality, a future in which wealthy hubs will concentrate all the wealth whereas some countries will go bankrupt or become data colonies. A powerful elite will rule a society in which large parts of the population will become useless, from an economic point of view. Harari asserts:

> Whereas in the past human had to struggle against exploitation, in the twenty-first century the really big struggle will be against irrelevance. And it is much worse to be irrelevant than exploited.[63]

Automation will thus produce people who are economically irrelevant, with no negotiation power over their existence, replaced by increasingly smart artificial intelligence. The 'Homo Economicus' will be the definition for the privileged few whereas the masses of workers will become 'Homo Intutilis'. Harari maintains:

> Consequently, despite the appearance of many new human jobs, we might nevertheless witness the rise of a new 'useless' class. We might actually get the worst of both worlds, suffering simultaneously from high unemployment and a shortage of skilled labour. Many people might share the fate not of nineteenth-century wagon drivers – who switched to driving taxis – but of nineteenth-century horses, who were increasingly pushed out of the job market altogether.[64]

Large parts of the population could thus face the destiny of nineteenth-century horses. Harari's example, from the early days of automobiles, can be interpreted not only as an example of how technology marginalizes a workforce. From a 21st-century perspective, with the knowledge of the intense environmental degradation and the pervasive impact on the cityscape caused by the introduction of automobiles, the question of the horses is much more multifaceted. The secondary effects of technological innovations were much more profound than car enthusiasts of the time could have ever imagined. As for the horses themselves: some of them found a new occupation as racehorses, or in agriculture, some became military horses, and some were released and could continue their lives as stray horses. Others found a different use, as Susanna Forrest describes, 'in the 1890s, a new American horse meat industry arose, if awkwardly. With the appearance of the electric street car and the battery-powered automobile, the era of the horse as a transportation technology was ending'.[65]

Becoming useless in economic terms can thus have very dangerous implications on livelihoods. Sofia Ghezzi sees the danger of polarization with one class in society remaining the only workforce, generating in collaboration with robots, AI, and other technical devices the wealth; and the rest, who will be largely without any revenue. Ghezzi asserts that 'the useless class is a threat to the peace of the community for two main reasons: i) they will need a stream of revenue to survive, and ii) they will need a new purpose in life'.[66]

The twofold challenge of automation is thus economic and ontological. How to assure economic survival for the excluded workforce and how to generate meaning as part of the leftover, useless class. In a society where economic productivity is the dominating measure for success and capital the dominating expression of this success, how to exclude a large section of people from the possibility of generating sources of revenue? The economic challenge has been addressed, largely through schemes of Universal Basic Income (UBI) or Universal Basic Services.

Rutger Bregman in his interpretation of a downsized utopia for realists sees the liberating potential of automation if combined with state intervention. Bregman writes: 'If we want to hold on to the blessings of technology, ultimately there's only one choice left, and that's redistribution. Massive redistribution. Redistribution of money (basic income), of time (a shorter working week), of taxation (on capital instead of labor), and, of course of robots'.[67]

The call for a robot tax also comes from Bill Gates who points out that if a worker has to pay income tax, then if 'a robot comes in to do the same thing, you'd think we'd tax the robot at a similar level'.[68] New taxes, such as robot tax, data tax, or taxes on digital services could thus generate new income to finance social imbalances created by automation. The first question is whether the new systems could potentially generate enough financial means to support large segments of the society; the second question is whether an existence on a basic level of economic sustainability provides enough sense of purpose and meaning to the 'useless class'.

As McKinsey Global Institute and other studies point out, the future members of the 'useless class' are not only to be found in the lower segments of the job market. Despite the fact that automation hits low-skilled labour in the first place, classic middle-class jobs are also increasingly endangered, such as mid-level management jobs in the fields of banking, insurance, the travel business, or in legal firms. The rise of precarity will thus enter the mainstream of society. As an OECD Study indicates, 'today the middle class looks increasingly like a boat in rocky waters'.[69] Even before the 4th Industrial Revolution (or the Second Machine Age) had started, the middle class in the OECD countries faced rising uncertainties, limited upward social mobility, higher risk of downward mobility, uncertain labour market perspectives, and a spiral of rising costs and stagnating incomes. As the report underlines 'economic insecurity concerns a large group of the population: more than one in three people are economically vulnerable, meaning they lack the liquid financial assets needed to maintain a living standard at the poverty level for at least three months'.[70] With one-in-six middle income jobs facing a high risk of automation, and more than one-in-five middle income households spending more than they earn, stability for the majority of the population appears to be in the realm of the unachievable. The question will be how to achieve political stability in a democracy when the economic fundaments of Western societies are increasingly crumbling? Zygmunt Bauman's assessment of the future possibilities is not optimistic. He writes: 'The volume of humans made redundant by capitalism's global triumph grows unstoppably and comes close now to exceeding the managerial capacity of the planet…'[71]

Especially in the European model, which since the 20th century is based on the co-existence of social welfare systems with what Joseph Stiglitz calls 'market fundamentalism',[72] the future redundancy of large parts of the (working) population will profoundly challenge the idea of social protection.

When Otto von Bismarck, Chancellor of the German Reich, looked at the 'social question' in Germany around 1880, he saw the risk of the impoverished classes radicalizing and becoming a danger to the stability and security of the Empire. The social welfare system he established in order to prevent social unrest was thus not an act of altruism but an acknowledgement that economic precarity on a mass scale is a threat to the status quo. The social protection the Bismarckian system introduced was unique in the world and found many imitators. Social innovation à la Bismarck was the introduction of a system that, based on social contributions by employers and salary-based contributions by employees, allowed for social security, steady wage employment, guaranteed pensions, health care and social benefits for the whole family. This system was gradually also adopted by Belgium, Luxembourg, Austria, and France. But, as some of the key features (steady wage employment, fixed retirement age, salary-based contributions) are fundamentally anchored in an industrial society with demographic growth and the need for workers, the Bismarckian model cannot function in a context in which the demographic pyramid is inverted, in which the need for a human workforce is drastically downsized and job stability, plannable careers, and salary growth have become wishful projections for an increasing number of employees.

In Gøsta Esping-Andersen's typology of welfare capitalism,[73] the Bismarkian model corresponds to the conservative regime. The other regimes Esping-Andersen maps out are the liberal regime and the social democratic regime. In Esping-Andersen's ideal, a social welfare system would guarantee social rights which permit people to make their living standards independent of pure market forces. She describes the ideal as a decommodification effect. Esping-Anderson sees a transformation of the role of the states, as she explains:

> What once were night-watchman states, law-and-order states, militarist states, or even repressive organs of totalitarian rule, are now institutions

> predominantly preoccupied with the production and distribution of social well-being'.[74]

If the conservative regime is exemplified by the Bismarckian model, the liberal type in which 'means-tested assistance, modest universal transfers, or modest social-insurance plans predominate',[75] is based on the ideas of Lord William Henry Beveridge. His eponymous report defined the British social welfare system after 1946. The Beveridge model is dominated by market-based approaches to alleviate social problems, modest insurances, and national health systems and a stigmatization of social welfare benefactors. Countries, aside from the UK, who have adopted this system are the United States, Canada, Ireland, and Australia. The limited social protection leads to higher levels of inequality in the societies that adopt the liberal regime. In addition, for Esping-Anderson, the decommodification effects are minimalized, which means that the whole society is fully dependent on market forces.

The third regime, the social democratic regime, is the dominant approach in the Scandinavian countries where equality is a guiding principle of social welfare. High taxes and high levels of social services and benefits ensure, according to Esping-Anderson, that the market forces are crowded out in favour of an essentially universal solidarity in favour of the welfare state, so that: 'All benefit; all are dependent; and all will presumably feel obliged to pay'.[76]

Recent studies have proposed to add new types of regimes to Esping-Anderson's three worlds of welfare capitalism; for example, Robert Urbé, who suggests that a distinct 'Mediterranean Model'[77] (countries like Portugal, Greece, and Spain with unequal development of all the branches of social insurance and a higher burden of social well-being on the family) or the 'Central and Eastern European' systems (which clusters together the former Warsaw block countries which are still in a formative period in relation to their social welfare systems).

The hybridized systems, such as Denmark's flexicurity or the Hartz reforms in Germany have already partially transformed what used to be social-democratic or conservative regimes. But regardless of their regime, overall, the potential rise of the 'useless class', in addition to the demographic decline in Europe, has the potential to pulverize social welfare in the way it was conceived in the last 150 years.

Bouget, Frazer, Marlier, and Sabato, in a study of national policies, have pointed out that, in the 21st century, welfare systems fulfil three functions:

> social investment, social protection and stabilisation of the economy. Indeed, the social investment approach strongly relies on the assumption that social and economic policies are mutually reinforcing and that the former, when framed in a social investment perspective, does represent a 'precondition' for future economic and employment growth. Social

> investment involves strengthening people's current and future capacities. In other words, as well as having immediate effects, social policies also have lasting impacts by offering economic and social returns over time, notably in terms of employment prospects or labour incomes.[78]

If formerly, the welfare state was either a palliative against the brutality of capitalism, a stabilization stronghold to keep the forces of communism at bay, or a Scandinavian Folkhemmet, then in the social investment paradigm it has to develop entrepreneurial strategies to mitigate the structural flaws of neoliberalism and contribute to innovation in an economic system that has run out of sustainable ways of producing growth.

The social innovation approach thus articulated would put the state into the position to 'reconvert' the 'useless class' in order to provide into a future employment perspective. In times of increasing precarity towards the future and a hyperdynamic job market, this would amount to planning an economy with a potentially high margin of error.

An alternative approach would be the universal basic income, in which the state disengages from the social investment perspective and merely provides a financial minimum in a post-work society.

The World Bank, in its World Development Report 2019, defines Universal Basic Income as a tool to build a guaranteed social minimum, with three design features:

> First, the program is aimed at every individual, independent of income or employment status. Second, participants do not have to fulfill any conditions or reciprocal co-responsibilities. Third, assistance is provided in the form of cash instead of in-kind transfers and services.[79]

Sigal Samuel examined projects in which conditional and unconditional basic incomes were introduced (either permanently or on an experimental base) and claims that '130 countries, from Mexico to Italy to Uganda, have instituted conditional cash transfers',[80] which range from supports for health checks or school attendance. Some programmes have linked unconditional basic income schemes to revenues from oil and gas, such as the Alaska Permanent Fund. Since 1976, 25% of Alaska's oil and mineral royalties are deposited into a fund from which every Alaska resident (children included) receives a yearly dividend. In 2022, the dividend was 3200 Dollars. In Iran, the government introduced an unconditional income scheme in 2010, giving out a monthly transfer that amounts to 29% of the median income. According to some estimates,[81] this, adjusted to the US median income scheme, would translate in the United States as a yearly extra payment of $16,389.64 to every citizen.

Since 1997, the Eastern Band of Cherokee Indians Casino dividend in North Carolina divides the casino revenues among the members of the tribe, allowing for a yearly unconditional income between $4000 and $6000.

The biggest European experiment of the introduction of the UBI was made in Finland between 2017 and 2019, when 2000 randomly selected unemployed people received an unconditional monthly transfer of 560 Euros. The experiment proved a failure when it came to the reinsertion of the unemployed into the job market but had positive effects on mental well-being. As the Social Insurance Institution, Kela, the Finnish government agency that ran the experiment states that the participants of the experiment reported that they 'were more satisfied with their lives and experienced less mental strain, depression, sadness and loneliness'.[82] The fact that the Finnish experiment was limited to unemployed people and implicitly targeted the reintroduction into the labour market is putting the 'unconditional' aspect of the income into question. Despite the discontinuation of the Finnish experiment, the UBI is currently debated under different labels in Italy as a citizen income or in Germany as 'Bürgergeld'.

As an alternative to UBI, the concept of Universal Basic Services (UBS) would give every citizen free access to education, health care, local transport, information, and shelter. As the UNESCO Inclusive Policy Lab points out: 'UBS is incremental, relatively cheap, and can achieve superior results to a system of unconditional cash payments alongside markets for commodified services'.[83]

The automation processes of the coming decades and their potentially devastating effects on the labour market thus have to be anticipated by a social welfare system that has outgrown its Bismarckian, Beveridgian, Scandinavian, or Mediteranian foundations. While social innovation approaches risk by demanding too much innovative capacity from the state, UBI/UBS approaches risk by being too undemanding to the citizens.

While the financial side of providing UBI or UBS could be solved by introducing robot taxes, data taxes, and financial transaction taxes, the question of whether an existence at a basic level of economic sustainability is meaningful for the new class of the socially excluded. Experiences in Germany with the recipients of 'Hartz 4' social welfare, have shown high levels of depression associated with long absences from the labour market.[84] On the other side of the spectrum, work, organized in a meritocratic neoliberal regime, produces high levels of burnout and exhaustion. According to the American Psychological Association's 2021 Work and Well-being Survey, 79% of employees have experienced work-related stress, and almost 60% report negative impacts such as lack of interest or motivation, 32% report emotional exhaustion, and 44% physical fatigue.[85] Thus, on the eve of the automation wave, for many people, work life produces stress, but longer periods of dependence on social welfare also produce stress. Joseph Stiglitz points out that 'the for-profit sector relies almost exclusively on extrinsic rewards',[86] but, ultimately, intrinsic motivation is gradually on the decline. At the same time, the feeling of redundancy experienced by more than one-third of German social welfare beneficiaries also hints at a problem of intrinsic motivation in the absence of extrinsic pressure.

An UBI might alleviate the financial pressures of the future 'useless class' but, locked in an existence on the minimum level with the luxury of free time allocation, the question of meaning will become more pressing. The crisis triangle – depopulated peripheries, new useless class, outdated social welfare systems – would thus have to be transformed into a space of possibility not only for economic but also for ontological reasons.

For those who are interested in an existence in which they can contribute to a higher goal than the satisfaction of their own desires, social welfare has to provide options and spaces. These spaces already exist, they are located in Abruzzo or Puglia, in Chemnitz or Liepaja, and in Sardinia or in Sicily. In an ironic twist of fate, the 'useless class' of capitalism could create in the 'ruined peripheral' spaces of capitalism a new way of organizing societies. These local societies could discover non-capitalist versions of political and economic freedom. The 'dreary flat' in Middlesex, Berlin Marzahn, or Seine-Saint-Denis could be traded for a house and a community in Castiglione di Sicilia, Pleven, or Liepaja. The answer could thus be: not universal basic income but peripheral basic space. Freedom from economic constraints combined with space and start-up capital to create new possibilities. In the age of future automation, the 'useless class' can thus become the 'socially innovative class' and, the peripheries the laboratories for new society arrangements.

Notes

1 Lewis Mumford, *The Story of Utopias* (New York: Boni and Liveright, 2008, original work published 1922), p. 14.
2 Ibid., p. 184.
3 Felix O. Murchadha, 'Being as Ruination: Heidegger, Simmel, and the Phenomenology of Ruins', *Philosophy Today*, 46 (2002), pp. 10–18 (p. 10).
4 Svetlana Boym, 'Tatlin or Ruinophilia', *Cabinet*, Issue 28/Bones, Winter (2007–2008) <https://www.cabinetmagazine.org/issues/28/boym2.php> [accessed 5 March 2022]
5 Rose Macaulay, *Pleasure of Ruins* (New York: Walker and Company, 1953), pp. xv–xvi.
6 Kate Abbey-Lambertz, 'Detroit's Abandoned Ruins Are Captivating, but Are They Bad for Neighborhoods?' (30 December 2013) <https://www.huffpost.com/entry/detroit-ruins_n_4519731> [accessed 6 March 2022]
7 Caitlin DeSilvey and Tim Edensor, 'Reckoning with Ruins', *Progress in Human Geography*, 37(4) (2012), pp. 465–485 (p. 465).
8 Ibid., p. 474.
9 Arata Isozaki and Maria F. González, 'The City in the Air', *ArchDaily* (18 March 2019) <https://www.archdaily.com/912738/the-city-in-the-air-by-arata-isozaki> [accessed 14 March 2022]
10 Robert Smithson, 'The Monuments of Passaic', *Artforum*, 6(4) (1967), p. 74.
11 Kaeleigh Herstad, 'Reclaiming Detroit', *The Public Historian*, 39(4) (2017), p. 91.
12 Detroit Future City, *139 Square Miles*, report by The John S. and James L. Knight Foundation (Detroit: Inland Press, 2017), p. 14.
13 Ibid., p. 19.

14 Ibid., p. 58.
15 Thomas Sugrue, *Origins of the Urban Crisis: Race and Inequality in Postwar Detroit*, rev. ed. (Princeton: Princeton University Press, 2005), p. 5.
16 John Patrick Leary, 'Detroitism: What Does 'Ruin Porn' Tell Us about the Motor City?', *Guernica* (15 January 2011) <https://www.guernicamag.com/leary_1_15_11/> [accessed 7 March 2022]
17 Ibid.
18 Robert Elmes, 'Why Galapagos is Moving to Detroit', *Galapagos Art Space* (2014) <https://www.galapagosdetroit.com/whydetroit> [accessed 7 March 2022]
19 J.C. Reindl, 'Entrepreneur Flips Building Near Detroit Train Station', *Detroit Free Press* (2019) <https://eu.freep.com/story/money/business/michigan/2019/01/09/galapagos-detroit-building-dan-gilbert/2524608002/> [accessed 7 March 2022]
20 Bellhop, '5 Up and Coming Neighborhoods in Detroit' (2021) <https://www.getbellhops.com/blog/5-up-and-coming-neighborhoods-in-detroit/> [accessed 7 March 2022]
21 Ibid.
22 Herstad, p. 101.
23 Leary (2011).
24 Michigan Central (2022) <https://michigancentral.com/about/> [accessed 7 March 2022]
25 Jefferson Cowie and Joseph Heathcott, *Beyond the Ruins: The Meanings of Deindustrialization* (Ithaca: ILR Press, 2003), p. 1.
26 Cowie and Heathcott, p. 15.
27 Detroit Future City, p. 33.
28 Camilo J. Vergara, *American Ruins* (New York: The Monacelli Press, 1999), p. 57.
29 Vince Carducci, 'On Art and Gentrification', *Infinite Mile* (January 2015) <http://infinitemiledetroit.com/On_Art_and_Gentrification.html> [accessed 15 March 2022]
30 Camilo J. Vergara, *Detroit Is No Dry Bones: The Eternal City of the Industrial Age* (Ann Arbor: University of Michigan Press, 2016), p. 281.
31 Jason Hackworth, 'Demolition as Urban Policy in the American Rust Belt', *Sage Journals,* 48(11) (2016) <https://doi.org/10.1177/0308518X16654914> [accessed 20 March 2022]
32 Leary (2011).
33 Anna Tsing, *The Mushroom at the End of the World* (Princeton: Princeton University Press, 2015), p. 18.
34 Ibid., p. 6.
35 Ibid.
36 DeSilvey and Edensor, p. 478.
37 Chemnitz in Zahlen, 'Einwohnerzahlen und –entwicklung', FOG-Institut für Markt- und Sozialforschung Chemnitz (2022) <https://www.chemnitz-in-zahlen.de/stadtteil-daten/einwohnerzahlen-und-entwicklung/> [accessed 5 August 2022]
38 Chemnitz Kulturhauptstadt Europas 2025, Zahlen und Fakten 'Bevölkerung (Stand: 31.07.2022)' <https://www.chemnitz.de/chemnitz/de/unsere-stadt/stadtportrait/zahlen-und-fakten/bevoelkerung.html> [accessed 5 August 2022]
39 Grotewohl, Otto, Speech 10 May 1953, Karl Marx Stadt <https://en.wikipedia.org/wiki/Chemnitz#cite_note-11> [accessed 14 March 2022]
40 Valentina Romei, 'In Charts, Europe's Demographic Time-Bomb', *Financial Times* (14 January 2020) <https://www.ft.com/content/49e1e106-0231-11ea-b7bc-f3fa4e77dd47> [accessed 14 March 2022]

41 Eurostat, 'Fertility Statistics', European Commission (April 2022) <https://ec.europa.eu/eurostat/statistics-explained/index.php?title=Fertility_statistics> [accessed 25 June 2022]
42 Romei (2020).
43 Dirk J. van de Kaa, 'Is the Second Demographic Transition a Useful Research Concept: Questions and Answers', in *Vienna Yearbook of Population Research*, Vol. 2 (Vienna: Austrian Academy of Sciences Press, 2004), pp. 4–10 (p. 7).
44 Ibid., p. 8.
45 Eurostat, 'Backround, Regions', European Commission (2021) <https://ec.europa.eu/eurostat/web/regions/background> [accessed 15 June 2022]
46 Eures, 'Short Overview of the Labour Market, Labour Market Information', *Europa-EURES* (2020) <https://ec.europa.eu/eures/printLMIText.jsp?lmiLang=en&regionId=BG3&catId=9574> [accessed 15 June 2022]
47 Eurostat, 'Backround, Regions' (2021).
48 AdminStat Latvia, 'Maps, Analysis and Statistics about the Resident Population', *Province of Liepaja* (2018) <https://ugeo.urbistat.com/AdminStat/en/lv/demografia/dati-sintesi/liepaja/170000/3> [accessed 15 June 2022]
49 Baltic News Network, 'CSP Reports Population Decline in Latvia' (1 June 2022) <https://bnn-news.com/csp-reports-population-decline-in-latvia-235142> [accessed 15 June 2022]
50 Eurostat, 'Statistics on Young People Neither in Employment nor in Education or Training', European Commission (2021) <https://ec.europa.eu/eurostat/statistics-explained/index.php?title=Statistics_on_young_people_neither_in_employment_nor_in_education_or_training> [accessed 15 June 2022]
51 Joseph Stiglitz, 'Moving beyond Market Fundamentalism to a More Balanced Economy', *Annals of Public and Cooperative Economics*, 80(3) (2009), pp. 345–360 (p. 351).
52 John M. Keynes, 'The Economic Possibilities for Our Grandchildren', in *Essays in Persuasion*, ed. by J.M. Keynes (New York: Harcourt Brace), pp. 358–373.
53 Kerstin Hötte, Melline Somers, and Angelos Theodorakopoulos, 'The Fear of Technology Driven Unemployment and its Empirical Base', Centre for Economic Policy Research (10 June 2022) <https://cepr.org/voxeu/columns/fear-technology-driven-unemployment-and-its-empirical-base> [accessed 20 July 2022]
54 Ibid.
55 McKinsey & Company, 'Jobs Lost, Jobs Gained: Workforce Transitions in a Time of Automation', McKinsey Global Institute (December 2017).
56 Euan Davis, '21 More Jobs of the Future: A Guide to Getting and Staying Employed through 2029', Cognizant (Teaneck: Center for the Future of Work, 2018) <https://www.cognizant.com/futureofwork/whitepaper/21-more-jobs-of-the-future-a-guide-to-getting-and-staying-employed-through-2029> [accessed 20 July 2022]
57 Ibid.
58 Ibid.
59 McKinsey & Company (2017), p. 8.
60 Ibid.
61 John Hawksworth, 'Is Artificial Intelligence Replacing Jobs? Here's the Truth', *World Economic Forum* (18 September 2018) <https://www.weforum.org/agenda/2018/09/is-artificial-intelligence-replacing-jobs-truth/> [accessed 20 July 2022]
62 Catherine Clifford, 'Elon Musk: Robots Will Take Your Jobs, Government Will Have to Pay Your Wage', *CNBC* (4 November 2016) <https://www.cnbc.com/2016/11/04/elon-musk-robots-will-take-your-jobs-government-will-have-to-pay-your-wage.html> [accessed 20 July 2022]

63 Yuval Noah Harari, 'Read Yuval Harari's Blistering Warning to Davos in Full', *World Economic Forum* (24 January 2020) <https://www.weforum.org/agenda/2020/01/yuval-hararis-warning-davos-speech-future-predications/> [accessed 20 July 2022]

64 Yuval Noah Harari, *21 Lessons for the 21st Century* (London: Jonathan Cape, 2018), p. 20.

65 Susanna Forrest, 'The Troubled History of Horse Meat in America', *The Atlantic* (8 June 2017) <https://www.theatlantic.com/technology/archive/2017/06/horse-meat/529665/> [accessed 20 July 2022]

66 Sofia Ghezzi, 'Artificial Intelligence Transforming Our Societal Structure - The Rise Useless Class', Institute for Internet & the Just Society (22 April 2022) <https://www.internetjustsociety.org/useless-class> [accessed 20 July 2022]

67 Rutger Bregman, *Utopia for Realists* (London: Bloomsbury, 2017), p. 199.

68 Arjun Kharpal, 'Bill Gates Wants to Tax Robots, but the EU Says, 'No Ways, No Way', *CNBC* (2 June 2017) <https://www.cnbc.com/2017/06/02/bill-gates-robot-tax-eu.html> [accessed 20 July 2022]

69 OECD, *Under Pressure: The Squeezed Middle Class* (Paris: OECD Publishing, 2019), p. 16, <https://doi.org/10.1787/689afed1-en>.

70 Ibid., p. 4.

71 Zygmunt Bauman, *Liquid Times: Living in an Age of Uncertainty* (Cambridge: Polity Press, 2007), p. 28.

72 Stiglitz, p. 347.

73 Gøsta Esping-Andersen, *The Three Worlds of Welfare Capitalism* (Cambridge: Polity Press, 1990).

74 Ibid., p. 1.

75 Ibid., p. 26.

76 Ibid., p. 28.

77 Robert Urbé, *Welfare in Europe, in The Future of the Welfare State: A Comparative Study in EU Countries* (Freiburg im Breisgau: Lambertus-Verlag, 2012), pp. 21–26.

78 Denis Bouget, Hugh Frazer, Eric Merlier Sebastiano Sabato and Bart Vanhercke, 'Social Investment in Europe – A Study of National Policies', European Commission (2015) <ESPN - Social Investment Synthesis Report _final.pdf>

79 World Bank, 'World Development Report: The Changing Nature of Work' (2019) <https://www.worldbank.org/en/publication/wdr2019> [accessed 25 March 2022]

80 Sigal Samuel, 'Everywhere Basic Income Has Been Tried in One Map', *Vox* (20 October 2020) <https://www.vox.com/future-perfect/2020/2/19/21112570/universal-basic-income-ubi-map> [accessed 25 March 2022]

81 Chris Weller, 'Iran Introduced a Basic Income Scheme, and Something Strange Happened', *World Economic Forum*, published in collaboration with *Business Insider* (31 May 2017) <https://www.weforum.org/agenda/2017/05/iran-introduced-a-basic-income-scheme-and-something-strange-happened> [accessed 25 March 2022]

82 Finnish Ministry of Social Affairs and Health, 'Results of the Basic Income Experiment: Small Employment Effects, Better Perceived Economic Security and Mental Wellbeing', Press Release (6 May 2020) <https://stm.fi/en/-/perustulokokeilun-tulokset-tyollisyysvaikutukset-vahaisia-toimeentulo-ja-psyykkinen-terveys-koettiin-paremmaksi> [accessed 25 March 2022]

83 Ian Gough, 'Move the Debate from Universal Basic Income to Universal Basic Services', UNESCO Inclusive Policy Lab (19 January 2021) <https://en.unesco.org/inclusivepolicylab/analytics/move-debate-universal-basic-income-universal-basic-services> [accessed 25 March 2022]

84 Maike Rademaker, 'Im Dschungel der Zuständigkeiten', *Deutschlandfunk* (26 December 2020) <https://www.deutschlandfunk.de/psychisch-kranke-im-hartz-iv-system-im-dschungel-der-100.html> [accessed 20 August 2022]
85 Ashley Abrasom, 'Burnout and Stress are Everywhere', *American Psychological Association*, 53(1) (2022) <https://www.apa.org/monitor/2022/01/special-burn-out-stress> [accessed 20 August 2022]
86 Stiglitz, p. 359.

Bibliography

Abbey-Lambertz, Kate, 'Detroit's Abandoned Ruins are Captivating, but are They Bad for Neighbourhoods?' (30 December 2013) <https://www.huffpost.com/entry/detroit-ruins_n_4519731> [accessed 6 March 2022]

Abrasom, Ashley, 'Burnout and Stress are Everywhere', *American Psychological Association*, 53(1) (2022) <https://www.apa.org/monitor/2022/01/special-burnout-stress> [accessed 20 August 2022]

AdminStat Latvia, 'Maps, Analysis and Statistics about the Resident Population', Province of Liepaja (2018) <https://ugeo.urbistat.com/AdminStat/en/lv/demografia/dati-sintesi/liepaja/170000/3> [accessed 15 June 2022]

Baltic News Network, 'CSP Reports Population Decline in Latvia' (1 June 2022) <https://bnn-news.com/csp-reports-population-decline-in-latvia-235142> [accessed 15 June 2022]

Bauman, Zygmunt, *Liquid Times: Living in an Age of Uncertainty* (Cambridge: Polity Press, 2007)

Bellhop, '5 Up and Coming Neighbourhoods in Detroit' (2021) <https://www.getbellhops.com/blog/5-up-and-coming-neighbourhoods-in-detroit/> [accessed 7 March 2022]

Bouget, Denis, Frazer, Hugh, Merlier, Eric, Sabato, Sebastiano and Vanhercke, Bart, 'Social Investment in Europe – A Study of National Policies', European Commission (2015) <ESPN - Social Investment Synthesis Report_final.pdf>

Boym, Svetlana, 'Tatlin or Ruinophilia', *Cabinet*, Issue 28/Bones, Winter (2007–2008) <https://www.cabinetmagazine.org/issues/28/boym2.php> [accessed 5 March 2022]

Bregman, Rutger, *Utopia for Realists* (London: Bloomsbury, 2017)

Carducci, Vince, 'On Art and Gentrification', *Infinite Mile* (January 2015) <http://infinitemiledetroit.com/On_Art_and_Gentrification.html> [accessed 15 March 2022]

Chemnitz in Zahlen, 'Einwohnerzahlen und –entwicklung', FOG-Institut für Markt- und Sozialforschung Chemnitz (2022) <https://www.chemnitz-in-zahlen.de/stadtteil-daten/einwohnerzahlen-und-entwicklung/> [accessed 5 August 2022]

Chemnitz Kulturhauptstadt Europas 2025, Zahlen und Fakten, 'Bevölkerung (Stand: 31.07.2022)' <https://www.chemnitz.de/chemnitz/de/unsere-stadt/stadtportrait/zahlen-und-fakten/bevoelkerung.html> [accessed 5 August 2022]

Clifford, Catherine, 'Elon Musk: Robots Will Take Your Jobs, Government Will Have to Pay Your Wage', *CNBC* (4 November 2016) <https://www.cnbc.com/2016/11/04/elon-musk-robots-will-take-your-jobs-government-will-have-to-pay-your-wage.html> [accessed 20 July 2022]

Cowie, Jefferson and Heathcott, Joseph, *Beyond the Ruins: The Meanings of Deindustrialization* (Ithaca: ILR Press, 2003)

Davis, Euan, '21 More Jobs of the Future: A Guide to Getting and Staying Employed through 2029', Cognizant (Teaneck: Center for the Future of Work, 2018) <https://www.cognizant.com/futureofwork/whitepaper/21-more-jobs-of-the-future-a-guide-to-getting-and-staying-employed-through-2029> [accessed 20 July 2022]

DeSilvey, Caitlin and Edensor, Tim, 'Reckoning with Ruins', *Progress in Human Geography*, 37(4) (2012), pp. 465–485

Detroit Future City, *139 Square Miles*, report by The John S. and James L. Knight Foundation (Detroit: Inland Press, 2017)

Elmes, Robert, 'Why Galapagos is Moving to Detroit', *Galapagos Art Space* (2014) <https://www.galapagosdetroit.com/whydetroit> [accessed 7 March 2022]

Esping-Andersen, Gøsta, *The Three Worlds of Welfare Capitalism* (Cambridge: Polity Press, 1990)

Eures, 'Short Overview of the Labour Market, Labour Market Information', Europa-EURES (2020) <https://ec.europa.eu/eures/printLMIText.jsp?lmiLang=en&regionId=BG3&catId=9574> [accessed 15 June 2022]

Eurostat, 'Background, Regions', European Commission (2021) <https://ec.europa.eu/eurostat/web/regions/background> [accessed 15 June 2022]

Eurostat, 'Fertility Statistics', European Commission (April 2022) <https://ec.europa.eu/eurostat/statistics-explained/index.php?title=Fertility_statistics> [accessed 25 June 2022]

Eurostat, 'Statistics on Young People Neither in Employment nor in Education or Training', European Commission (2021) <https://ec.europa.eu/eurostat/statistics-explained/index.php?title=Statistics_on_young_people_neither_in_employment_nor_in_education_or_training> [accessed 15 June 2022]

Finnish Ministry of Social Affairs and Health, 'Results of the Basic Income Experiment: Small Employment Effects, Better Perceived Economic Security and Mental Wellbeing', Press Release (6 May 2020) <https://stm.fi/en/-/perustulokokeilun-tulokset-tyollisyysvaikutukset-vahaisia-toimeentulo-ja-psyykkinen-terveys-koettiin-paremmaksi> [accessed 25 March 2022]

Forrest, Susanna, 'The Troubled History of Horse Meat in America', *The Atlantic* (8 June 2017) <https://www.theatlantic.com/technology/archive/2017/06/horse-meat/529665/> [accessed 20 July 2022]

Ghezzi, Sofia, 'Artificial Intelligence Transforming Our Societal Structure - The Rise Useless Class', Institute for Internet & the Just Society (22 April 2022) <https://www.internetjustsociety.org/useless-class> [accessed 20 July 2022]

Gough, Ian, 'Move the Debate from Universal Basic Income to Universal Basic Services', UNESCO Inclusive Policy Lab (19 January 2021) <https://en.unesco.org/inclusivepolicylab/analytics/move-debate-universal-basic-income-universal-basic-services> [accessed 25 March 2022]

Grotewohl, Otto, Speech 10 May 1953, Karl Marx Stadt <https://en.wikipedia.org/wiki/Chemnitz#cite_note-11> [accessed 14 March 2022]

Hackworth, Jason, 'Demolition as Urban Policy in the American Rust Belt', *Sage Journals*, 48(11) (2016) <https://doi.org/10.1177/0308518X16654914> [accessed 20 March 2022]

Hacksworth, John, 'Is Artificial Intelligence Replacing Jobs? Here's the Truth', *World Economic Forum* (18 September 2018) <https://www.weforum.org/agenda/2018/09/is-artificial-intelligence-replacing-jobs-truth/> [accessed 20 July 2022]

Harari, Yuval Noah, *21 Lessons for the 21st Century* (London: Jonathan Cape, 2018)

Harari, Yuval Noah, 'Read Yuval Harari's Blistering Warning to Davos in Full', *World Economic Forum* (24 January 2020) <https://www.weforum.org/agenda/2020/01/yuval-hararis-warning-davos-speech-future-predications/> [accessed 20 July 2022]

Herstad, Kaeleig, 'Reclaiming Detroit', *The Public Historian*, 39(4) (2017), pp. 85–113

Hötte, Kerstin, Somers, Melline & Theodorakopoulos, 'The Fear of Technology Driven Unemployment and its Empirical Base', Centre for Economic Policy Research (10 June 2022) <https://cepr.org/voxeu/columns/fear-technology-driven-unemployment-and-its-empirical-base> [accessed 20 July 2022]

Isozaki, Arata and González, Maria F., 'The City in the Air', *ArchDaily* (18 March 2019) <https://www.archdaily.com/912738/the-city-in-the-air-by-arata-isozaki> [accessed 14 March 2022]

Keynes, John M., 'The Economic Possibilities for our Grandchildren', in *Essays in Persuasion*, ed. by J.M. Keynes (New York: Harcourt Brace, 1963)

Kharpal, Arjun, 'Bill Gates Wants to Tax Robots, but the EU Says, 'No Ways, No Way', *CNBC*, 2 June 2017 <https://www.cnbc.com/2017/06/02/bill-gates-robot-tax-eu.html> [accessed 20 July 2022]

Leary, John Patrick, 'Detroitism: What Does 'Ruin Porn' Tell Us about the Motor City?', *Guernica* (15 January 2011) <https://www.guernicamag.com/leary_1_15_11/> [accessed 7 March 2022]

Macaulay, Rose, *Pleasure of Ruins* (New York: Walker and Company, 1953)

McKinsey & Company, 'Jobs Lost, Jobs Gained: Workforce Transitions in a Time of Automation', McKinsey Global Institute (December 2017)

Michigan Central (2022) <https://michigancentral.com/about/> [accessed 7 March 2022]

Mumford, Lewis, *The Story of Utopias* (New York: Boni and Liveright, 2008, original work published 1922)

Murchadha, Felix O., 'Being as Ruination: Heidegger, Simmel, and the Phenomenology of Ruins', *Philosophy Today*, 46, January (2002)

OECD, *Under Pressure: The Squeezed Middle Class* (Paris: OECD Publishing, 2019). <https://doi.org/10.1787/689afed1-en>

Rademaker, Maike, 'Im Dschungel der Zuständigkeiten', *Deutschlandfunk* (26 December 2020) <https://www.deutschlandfunk.de/psychisch-kranke-im-hartz-iv-system-im-dschungel-der-100.html> [accessed 20 August 2022]

Reindl, J.C., 'Entrepreneur Flips Building near Detroit Train Station', *Detroit Free Press* (2019) <https://eu.freep.com/story/money/business/michigan/2019/01/09/galapagos-detroit-building-dan-gilbert/2524608002/> [accessed 7 March 2022]

Romei, Valentina, 'In Charts, Europe's Demographic Time-Bomb', *Financial Times* (14 January 2020) <https://www.ft.com/content/49e1e106-0231-11ea-b7bc-f3fa4e77dd47> [accessed 14 March 2022]

Samuel, Sigal, 'Everywhere Basic Income Has Been Tried in One Map', *Vox*, (20 October 2020) <https://www.vox.com/future-perfect/2020/2/19/21112570/universal-basic-income-ubi-map> [accessed 25 March 2022]

Smithson, Robert, 'The Monuments of Passaic', *Artforum*, 6(4) (December 1967)

Stiglitz, Joseph, 'Moving beyond Market Fundamentalism to a More Balanced Economy', *Annals of Public and Cooperative Economics*, 80(3) (2009)

Sugrue, Thomas, *Origins of the Urban Crisis: Race and Inequality in Postwar Detroit*, rev. ed. (Princeton: Princeton University Press, 2005)

Tsing, Anna, *The Mushroom at the End of the World* (Princeton: Princeton University Press, 2015)

Urbé, Robert, 'Welfare in Europe', in *The Future of the Welfare State: A Comparative Study in EU Countries* (Freiburg im Breisgau: Lambertus-Verlag, 2012)

van de Kaa, Dirk J., 'Is the Second Demographic Transition a Useful Research Concept: Questions and Answers', in *Vienna Yearbook of Population Research*, Vol. 2 (Vienna: Austrian Academy of Sciences Press, 2004)

Vergara, Camilo J., *American Ruins* (New York: The Monacelli Press, 1999)

Vergara, Camilo J., *Detroit Is No Dry Bones: The Eternal City of the Industrial Age* (Ann Arbor: University of Michigan Press, 2016)

Weller, Chris, 'Iran Introduced a Basic Income Scheme, and Something Strange Happened', in *World Economic Forum*, published in collaboration with *Business Insider* (31 May 2017) <https://www.weforum.org/agenda/2017/05/iran-introduced-a-basic-income-scheme-and-something-strange-happened> [accessed 25 March 2022]

World Bank, 'World Development Report: The Changing Nature of Work' (2019) <https://www.worldbank.org/en/publication/wdr2019> [accessed 25 March 2022]

5 Retopia as an Innovation System

A Redefinition of Innovation

The power of the There Is No Alternative (TINA) doctrine is perhaps implicit in its claim that any alternatives that are articulated have to be foolproof against criticism before they can be taken into consideration. No alternative can live up to such a high standard.

The irony is that the biggest library in the world, the Library of Congress in Washington D.C., could be filled with books, articles, TV programmes, and films that are articulating a fierce critique of neoliberalism. Probably the 1350 kilometers of shelf space and the capacity to contain 164 million items would scarcely be sufficient to accommodate all the works that are pointing out the flaws of (neo)liberal politics and economics.

Apart from the obvious criticisms against its destructive effects on the lives of billions, neoliberalism's enclosure of the mind scape is logically incoherent. The TINA doctrine is contradictory to the self-proclaimed values of neoliberalism, especially with its claim on freedom (Friedman's 'whatever they value') and dynamism. If the freedom is limited to economic freedom and implicitly excludes any collective alternatives, and the dynamism is locked in a mind-set that cements the existing arrangements, how can neoliberalism deliver on its premise of enhancing the freedom and creating dynamic societies?

It is thus fair to acknowledge, in advance, that what will follow, the proposal for a new political imaginary, is certainly not a perfect, flawless project but a contribution to open up the debate and encourage others to participate in the process of prefiguring alternative society arrangements. The flaws of neoliberal, converted, apocalyptic and nostalgic, and imaginaries certainly provide a good motivation to attempt to scheme up possible alternatives. Thus far, this is a process of (individual) political imagination that will remain without consequences, if it does not appeal to any relevant group interested in the experiment.

Experiment is derived from the Latin 'experiti', so 'try', which implies that there is no guarantee for success. A political project based on an experiment is thus an attempt to do something new. In economic terms every

DOI: 10.4324/9780429288180-5

entrepreneur who is trying to establish a new business is therefore experimenting. Economic experimentation is generally connoted in a positive way. Across the globe hundred-thousands of start-ups are often state supported, receive positive media coverage, and are the incorporation of dynamism, pioneer spirit, and hope for a better (economic) future. In political terms, anybody who is trying to establish a new imaginary is more likely to be seen, at best, as a hopeless dreamer and, at worst, as a fool.

But, as Geoff Mulgan has emphasized, social innovation is based on experimentation. Mulgan asserts 'no one knows exactly which models will work in the future. The only way to find out is to experiment and learn quickly.'[1] One possible outcome of every experiment is failure, but, according to Mulgan, 'it's only by testing things out on a small scale that society is able to learn about what works and what could change.'[2]

An experimental approach is thus an appropriate tool in an era of growing complexity. As Smedt, Borch, and Fuller emphasize:

> Innovation in the 21st century differs from the model embraced in the last century (i.e. profit-oriented and nationally targeted) with a linear, technological and deterministic characteristic [...] Our understanding of the unpredictable inter-dependency of multiple forces means that more experimental approaches to creating new solutions are necessary. Thus, it is through trial-and-errors and learning-by-doing experiments that new solutions can be found to grand challenges.[3]

So therefore, at the outset, experimenting in the political realm is becoming a pre-requisite for social innovation. In the impasse of neoliberalism. it is foolish not to experiment with different set-ups of society. Projects that are experimenting with new forms of political arrangements are the precondition to new knowledge. Developing political projects means developing images. The Latin roots of 'project' are 'pro' (forth), and 'jacare' (throw/cast). Any project is thus per definition a way of casting an image ahead. It is time to develop new imaginaries that anticipate different futures. As Michel wrote:

> Prospective is above all an attitude of mind (anticipation and will) and a way of behaving (with imagination and hope) mobilized in order to ensure quality and control of our present and future existence. Prospective reinstates desire as the productive force of the future. If it has no future direction the present is empty of meaning. Thus dreams, do not oppose reality, they create it, and a plan animated by desire is the driving force for action.[4]

From a normative point, any alternative political imaginary should be motivated by Ernst Bloch's dictum that 'the defeated man must try the outside world again'[5] (in the aftermath of the Covid-related lockdowns, this has taken on a new connotation). The dominant emotional drive of hope (in the understanding of hope for society, not privatized hope) is thus a key element for any project that aims at innovation in the future. The spatiotemporal scope of such imaginaries would be, due to its experimental nature, local (the lesson of the 20th century was that grand-scale society experiments result in authoritarianism and totalitarianism) and located not in an abstract future but in the concrete and immediate future. The agency of the imaginary is based on the idea that societies are (wo)man made constructions. Thus, socially defined answers to the scarcity gap can be designed on a local level. The experiment-based imaginary is thus an attempt to open up the political horizon which is currently narrowed down by the four political imaginaries (neoliberal, converted, apocalyptic, nostalgic) described in chapter 1.

A new imaginary would also produce new stories. The premise, as outlined in Chapter 1 would be: a group of people from all over the world settles in a (half-) abandoned village in the periphery and designs a world according to their values. The battle of the story of this retopian story, animated by a retopian, experiment-based, imaginary, would be: it is possible to not only to do business start-ups, but also society start-ups. To come back to the five elements of narrative power analysis, the impact of stories can be understood by conflict, character, imaginary, foreshadowing and assumptions. The retopian story attempts to challenge the status quo. The conflict is defined by the concrete struggle to improve the conditions of living in an environment that is peripheral. The protagonists have to redefine livelihood outside of the realm of market fundamentalism and find new arrangements of common coexistence. The characters are pioneers who are leaving old ways of living behind in order to engage in an open-ended adventure. The imagery is a (half-) ruined place that gradually comes back to life. The retopian story foreshadows the idea that change is possible. Its assumption is that humans can transform from 'useless class' to 'social innovator'.

The retopian story fills the myth gap of the TINA doctrine, aiming to shape a different future. From neoliberal inadequacy (for those who are deemed useless) to empowerment the retopian quest narrative projects the possibility of inner and outer change. It delivers an explanation: if you are deemed useless for capitalist value creation, define your 'usefulness' in a new context. And the meaning of the story: humans have the potential to find better ways of living. The protagonists of the story would thus be active, pursue their desire, struggle against the external forces, pursue a difficult goal and internally change as a result. No more discussions about world revolution, like in the Marxist story, no more belief in the naïve 'rags to riches' plots of the Friedmanian story, no more passive 'I would prefer not to' protagonists like Bartleby, no more rejection of stories like in the Matsutake premise and no more transformation into critters in the chthulucene. The retopian story has a location (the ruined periphery), multiple protagonists (adventurers who want to revitalize deserted spaces), a premise (action and experiment lead to new ways of living), a shared story based on a shared location and a shared, immediate, experience.

The possibility to transform land in the periphery, to use spaces of possibilities in the interstices and to locally define new arrangements in governance, economic system, and social coexistence are at the core of the reconstructive projects, the retopias. Retopia is thus a return to Lewis Mumford's ideas of eutopia in a 21st-century environment defined by the need for transformative solutions in the face of environmental degradation, technical unemployment, demographic shifts, and geographical imbalances.

Mumford's assessment that the scope of reconstructive projects is local, located in the places deserted by capitalism and the possibility of practical execution is high is highly relevant today. Due to the global scope of the environmental and economic crisis, the possibility for political change is perceived impossible or realizable only in the far future. Mumford claimed that the notion that 'no effective change can be brought about in society until millions of people have deliberated upon it and willed it is one of the rationalizations which are dear to the lazy and ineffectual.'[6] The local level and the immediate level are thus the grounds on which change can be effectuated.

David Harvey defines spatial fix as 'capitalism's insatiable drive to resolve its inner crisis tendencies by geographical expansion and geographical restructuring.'[7] This insatiable drive for expansion has left an increasing number of regions behind. They are 'liberated' from value creation. Often these regions are left in some ruined state of the other, with high levels of poverty, lack of infrastructure and a spirit of defeat. In an inverted 'spatial fix' with the intention to restructure the ruined spaces, these places can become places where political freedom and new approaches to economy are possible again. These places are the topoi for retopias.

Retopias are multiple, practice-led collective projects, communities of practice that experiment with different forms of society in different 'ruined' (henceforth replaced with peripheral, in the sense of peripheral for the capitalist value-creation) regions of the world. The local retopian projects are completely free in their definition of the good society so that each retopian project is a laboratory of society.

In *The Story of Utopias* (1922), Mumford, puts up the nine criteria of what he considers to be the 'foundation of eutopia'[8]:

1. no single utopia for a single unit called humanity;
2. no single stratification (such as working class) as foundation for eutopia;
3. a utopia should be based on local habitation not on the idea of a nation state;
4. every community should have territorial limits;
5. every community should have the legislation that reflects the local conditions;

6. inhabitants of eutopia have to be familiar with local environment and its resources which means that the focus should be on use of local resources;
7. cultivated life is essentially a settled life therefore diversity and stable roots are important;
8. inhabitants of eutopia would have to accept life in a more limited environment than that of the great metropolitan centers, but their mental environment will be open to outer influences and not limited by national culture;
9. the eutopian culture is a culture of the environment that refuses to exploit other people's environment.

Mumford's nine criteria for eutopia can be transferred to the idea of retopia. Open, local, focused on local resources, diversity and stable roots, a renouncement of the lifestyle of the overpriced metropolitan centers, and a respect for other people's environment are also good foundations for retopia. Today, the networking possibilities create means to interlink different projects in different locations, to ensure know-how transfer, communication, and an open mental environment.

Retopia is thus a political imaginary aimed at defining spaces of possibility that are located in the periphery of capitalist value creation. In neoliberal language, these could be labelled as 'society start-ups'; in experimental terms, 'laboratories of societies'; and in social terms, spaces to 'try the outside world again'. They are an explicit appeal to replace protest and criticism with an active spirit directed towards practical improvement and discovery.

The example of the American pioneers going westwards is tainted with the mass murder and brutal displacement of the native population. But Retopia is an appeal to a pioneer spirit that is nonviolent and not aimed at expropriating the local population but rather at revitalizing spaces that are on a steep decline.

Retopia as a Method

In the contemporary context of the degradation of the economic, ecological, social, and political, conditions for new imaginaries cannot be found on islands like Utopia. There are no blueprints and no paradises to take inspiration from. The retopian imaginaries have to develop on the grounds of ruins, and the solutions are based on practical political action and reconstruction of the peripheries. Utopia is located on an idealized island, Retopia is located on the edge.

In order to open the prospective horizons and influence the 'not yet being', in the Blochian sense, in a positive way, utopia as a method (and per extension, retopia as a method) has to avoid the trap of creating narrow blueprint utopias. The complex realities of the 21st century are incompatible with generalizations, determinisms, and projections of limited private value systems on the whole of humanity.

Zygmunt Bauman addresses the most pertinent issue: 'We need to brace ourselves for a long period marked by more questions than answers and more problems than solutions, as well as for acting in the shadow of finely balanced chances of success and defeat'.[9] What Bloch called in the 1950s 'the darkness of the lived moment', in times of transition between world systems and interregnum, is thus tainted with even more shadows than at the time of Bloch's diagnosis.

According to Levitas, there is no need to proclaim any human essence, or human nature, but to acknowledge the scarcity gap and recognize that any assumption of how to overcome the scarcity gap by necessity indirectly implies preferences for the development of society. To state it otherwise: choices need to be made about how the scarcity gap can be overcome and these choices necessarily include normative preferences but are by no means an essentialist statement about human nature. The current liberal-democratic order is also based on implicit judgements about human nature, as Goodwin and Taylor point out:

> In fact, much liberal-democratic criticism of utopian thinking can be reduced to certain tenets about human nature: that men will always be as competitive and avaricious as capitalism makes them, and as corruptible by power. Ultimately, the evaluation of utopianism depends on whether that belief can be suspended.[10]

Goodwin and Taylor introduce 'belief' into the equation when they speak about models of society. The suspension of disbelief is an essential feature in storytelling. In political storytelling, the narrative fundament of any imaginary, the question of human nature is often addressed in reductive ways that tell more about the value of the observer. If a belief becomes too narrow and too reductive, the danger of fundamentalism always looms around the corner. The neoliberal imaginary with its implicitly reductive claim on human nature is thus reducing the scope of human agency.

The neoliberal fundamentalism (and its implicit claim on human nature) thus narrows down the discussion about solutions to the scarcity gap. The rational choice assumption of the *homo economicus* and the reduction on self-interest are reducing the potential of finding new political and economic arrangements to define human coexistence.

In this understanding, utopias serve as 'suggestive devices which help us decide in which directions change should take place'.[11] In the same vein, retopia can be understood as a necessary suggestive device that openly articulates ways in which society can develop without claiming certainty about having absolute knowledge about the result this transformation would bring. In the same way as it is impossible to know the outcome of the current economic system, it is impossible to offer certainty that a utopian alternative will result in a better society. Or, as Immanuel Wallerstein described the situation when he was interviewed by the artists Molly Nesbit, Hans-Ulrich Obrist, and Rikrit Tiravanija:

> you have to act the way you think you ought to act without the certainty that you will be successful. And that is very tricky. Lots of people want to be mobilized by certainty, but I think that is just not possible.[12]

De Jouvenel emphasizes the open-endedness of the future when he writes that the future 'is not the realm of the "true and false" but the realm of "possibles"'.[13] Consequently, he also rejects a science of the future in favor of a thinking of the future. Truth claims in relation to the 'not yet' have no scientific foundation and are thus in the realm of speculation. They are necessary intellectual and scientific endeavours but they do not offer the comfort of epistemological certainty. But the contemporary climate of crisis (which is also a crisis of knowledge) abstaining from the reflection of the 'possibles' is the equivalent of an intellectual capitulation. In this sense, the epistemological certainty that utopian alternatives are impossible would amount to totalitarian thinking. As Neurath observed already in 1919:

> To characterize utopias as representations of impossible events is altogether unfounded; one can in fact basically never predict whether a contemplated regime will at some point and some place become a reality.[14]

The speculative epistemology is obviously also a normative one, as to what kind of knowledge is accepted as scientific. Lewis Mumford (1922) asserts:

> Indeed, scientific knowledge has not merely heightened the possibility of life in the modern world: it has lowered its depths. When science is not touched by a sense of value it works – as it fairly consistently has worked during the past century – towards a complete dehumanization of the social order.[15]

Mumford criticizes that science has lowered the depth of human possibilities by its dissociation from art, dreaming, and emotional activities. An epistemology that excludes the imagination is impaired and static in the scope of knowledge that it values and produces.

In retopian imaginaries, the epistemological foundation is thus open, including the realm of the imagination, the world of the 'possibles'. Such an openness is only possible when any claim on absolute knowledge vanishes from the horizon.

The utopian method is an attempt to free the imagination from its enclosure. Utopian methodology, per definition, cannot be based on finding facts because utopian thinking is to a large degree counterfactual. Unlike postfactual thinking, which is based on circumventing truth claims in order to legitimize (reactionary) political agendas, counterfactual thinking is an act of imagination that corresponds to de Jouvenel's idea of the future, so 'what shall come into being'. It is based on an open discussion of possible futures.

Ruth Levitas proposes what she calls the Imaginary Reconstitution of Society as a method which has three modes: the archaeological mode, the architectural mode, and the ontological mode. The archaeological mode consists of an analysis of 'the underpinning model of the good society'[16] – open to scrutiny and thereby to public critique. The taxonomy of the contemporary imaginaries in Chapter 1 is an example of an analysis in the archaeological mode suggested by Levitas. The architectural mode corresponds to holistic modelling of a better society and exposes the danger of flawed utopias if the outcome is a blueprint utopia. The retopian imaginary is an attempt to avoid blueprint modelling. The ontological mode corresponds to a future in which the ideals of the new society have modified the arrangements of living. The stories from an open future in Chapter 6 are a short fictional exercise in sketching such an ontological mode. Per nature, such an exercise is in the realm of speculation.

What is the function of a utopian reconstruction of societies? As Levitas recognizes 'to focus on the function of utopia is already to move away from the colloquial usage, which says nothing about what utopia is for, but implies that it is useless'.[17]

While there are good reasons to recognize that utopias can have a negative impact, Levitas opts for acknowledging the positive function of utopias because of their potential to combat fatalism and their capacity to express better ways of being. Levitas thus asserts functions of utopia: 'Three different functions, not necessarily mutually exclusive, are thus identified: compensation, critique, catalyst'.[18]

The compensation function is synonymous with escape – Lewis Mumford (1922) would call this a 'utopia of escape'[19] – and has no political intention but is a mere idle dream of a different life. The critical function is corresponding to More's approach in his foundational text *Utopia*: the island of Utopia, as an ideal commonwealth, serves as an ideal which illustrates the shortcomings of the existing society. The catalyst function explicitly aims at implementing a new model of society. Retopian imaginaries thus have a catalyst function, through their practice-based and experimental approach and through the diversity manifested in the laboratories of societies.

The Imaginary Reconstitution of Society is thus a method to describe approaches to utopia in its constitutive modes and functions with the aim to systematically articulate the visions of the future and their potential for implementation.

Another methodology in the field of utopian thinking (although the authors meticulously avoid the word 'utopia') has been developed by Haiven and Khasnabish (2014). They define what they call prefigurative methodology as 'a methodology borrowed from the future we should like to see created, brought back into the present to help make the future a possibility'.[20]

Haiven and Khasnabish assert that 'the imagination is not some transcendental spirit but an embodied presence'.[21] Therefore, individual imaginations are constrained by specific circumstance and experiences. Shared imaginaries are thus creating the possibility to transcend the narrow horizon of personal experience and enter the process of co-creating new imaginaries. These shared imaginaries can also transcend the privatization of hope. As Haiven and Khasnabish underline: 'It represents our capacity to imagine and make common cause with the experiences of other people; it undergirds our capacity to build solidarity across boundaries and borders, real or imagined'.[22] For Haiven and Khasnabish, refraining from developing shared imaginaries means that 'we are left only with the residual dreams of the powerful and for the vast majority they are not experienced as dreams but as nightmares of insecurity, precarity, violence, and hopelessness'.[23]

The prefigurative methodology thus also has a normative dimension (collective imagination is transcending the horizon, private imagination is co-opted by the 'powerful'). The future ideal is thus projected back into the present in order to fuel the development of shared imaginaries that are, in turn, improving the possibility for collective action. In Haiven and Khasnabish's insistence on projecting positive images into the future (and positive images from the future back into the present) Fred Pollack's influence is perceivable. The move from diagnosis to prognosis, the insistence on positive future images as a self-reflection of the values, the possibility of the radical reversal of the order and the insistence that the creation of future images echoes Pollack's book *The Future Image*. Pollack wrote:

> The choice for modem man is no longer between this image of the future and that, but between images of his own choosing and images which are forced upon him by outside pressures. The empty void between today and tomorrow cannot withstand the magnetic pull of tomorrow...[24]

Renouncing to define future images thus means renouncing control over the future. A method of prefiguration or imaginary reconstitution of society constitutes an essential element of political imaginaries. The method is helping to articulate the function of the utopian vision (compensation, critique, catalyst), to identify the collective horizon and to project future images.

Haiven's and Khasnabish's prefigurative methodology are also part of what De Smedt, Borch, and Fuller classify as a 'visionary mode of thinking'[25] (as opposed to the eventualities mode of thinking and the predictive mode of thinking) since it focuses on preferable futures and envisions how society can be designed in a better way. The underlying normative dimension of the imaginaries is thus an important aspect of the mode of thinking. De Smedt, Borch, and Fuller said that 'What will happen?' scenarios lead to prediction. But they underline that:

'What can happen?' scenarios are normative scenarios - concerned with achieving particular future objectives - which lead to preserving and transforming scenarios. Preserving scenarios are used when the target can be met within an existing structure, while Transforming scenarios feature a form of backcasting, asking what would need to be changed for the target futures to be achieved.[26]

Retopias, in the current pre-experimentation stage are thus transforming scenarios which feature a form of backcasting. The starting point of backcasting is the definition of an idea of a desirable future. Then, from the point of view of the desirable future, the policies and programs should be identified that will connect the future to the present. This allows exploration of the preconditions that could lead to the desirable future and the definitions of strategies to reach the desired outcome.

In the retopian scenario, the desirable outcome is the revitalization of peripheral 'ruined' spaces and the reactivation of people who are either tired of the current neoliberal arrangements in society or who are made redundant by economic restructuration processes or future developments of automation. From a strategic point of view, the question is how to connect the spaces of possibilities, the ruined periphery, with the groups of people who are actively interested in developing new arrangements of society and willing to contribute actively to their realization. As Chapter 3 has shown, the current versions of intentional communities have appeal to only a limited group of people. Retopias, in order to be effective, need a more diverse targeting, a better strategic co-operation with the public sector, and a better balance of community and individual freedom. Retopia has to be conceived as a social innovation system.

The Quest Plot in the Ruined Periphery

A group of 100 people develop a project for a half-deserted village in Sicily; in exchange, they get empty houses and a development fund. In agreement with the remaining residents of the village, they decide on a plan to develop the village, create their own system of governance, and define the economic relationships and the local societal arrangements autonomously. The village gets 100 new inhabitants who, in exchange, get spaces and dedicated funds to develop their individual and collective ideas of a better life. The basic setup of retopia is the setup for a social innovation scheme.

Despite the fact that the concept of social innovation is a concept that has animated academic discussions for more than a decade, no generally accepted definition of the concept exists. In the context of territorial development, Frank Moulaert and Pieter Van den Broeck have underlined that: 'Spaces and places as habitats of hope and change are a very important focus in social innovation action research today'.[27] Moulart and Van den Broeck identify two contradictory ideologies in the definition of social innovation today: 'caring liberalism' and 'sociopolitical transformative social innovation'. The caring liberalism approach is focusing on making the market more inclusive and integrating social economy initiatives into the existing market structures through the provision of institutional spaces and social economic organizations. The sociopolitical transformative social innovation considers social innovation as a:

> strategy and process not only to satisfy individual and collective needs abused by the market, but to strengthen the solidarity content of social relations between people involved in social innovation initiatives, as well as call up these relations as triggers of socio-political empowerment'.[28]

In relation to creating spaces as habitats of hope, the two important aspects are the strengthening of the solidarity and the socio-political empowerment. Retopia as an innovation aims to provide spaces of possibility to groups of people with social innovation projects. Through the reactivation of peripheries and the creation of a participatory for a new framework for livelihood, solidarity and socio-political empowerment are stimulated.

Pisano, Lange & Berger claim that there are five key elements at the core of social innovation:

> 1) Novelty – social innovation should be new in some way, either new to the field, sector, region, market or user, or to be applied in a new way;
> 2) Implementation–social innovation is concerned with the practical application or implementation of a new idea that needs to be (or have the potential to be) financially sustainable in the mid- to long-term;
> 3) Meets a social need: One of the distinguishing features of a social innovation is that it is explicitly designed to meet a social need, understood as something that can cause serious harm or socially recognisable suffering when not met;
> 4) Effectiveness–Social innovation should be more effective than existing solutions by creating a measurable improvement in terms of outcomes [...]
> 5) Enhances society's capacity to act: The process of social innovation is important: in fact, the process of social innovation enhances society's capacity to act and often entails changes in social and power relations. It empowers beneficiaries by creating new roles and relationships, developing assets and capabilities and/or better use of assets and resources.[29]

Retopian projects connect new communities with peripheral regions. Land and housing are allocated to project-based groups which can give new scope and shape to the peripheral land. The social welfare system is partially reorganized in order to provide more autonomous forms of society arrangements for those who are becoming or risk becoming part of the new 'useless class' in times of automation. State subsidies are reconverted into social innovation investment in order to provide financial sustainability (or 'social start-up capital') to the establishing communities. The retopian approach is new, has the potential for financial self-sustainability in the mid-term, focuses on the social needs of the periphery and the social innovators, and has the potential to be more effective than the traditional welfare strategies or regional development strategies. In addition, it has a high potential to activate the periphery. All the beneficiaries would have the possibility to establish new roles and relationships and develop new assets and capabilities.

Geoff Mulgan calls for an increased state investment in social innovation initiatives. He defines social innovations as projects that 'can be defined and potentially spread beyond their initial context'.[30] The initiatives should be provided by organizations, meet socially recognized needs, and work in circumstances where markets have failed.

It is quite clear that markets have failed to function in regions which are economically undeveloped, whether Liepaja, Pleven, or Castiglione di Sicilia, and if the dire prognostics of the labor market in times of future automation turn into a reality, the market will also fail to provide enough employment in all strata of society. Employment and regional development are thus not only economic needs but also social needs. Thus organizations, in this case groups that are developing projects to develop the region, become stakeholders by the mere fact that they are active participants and residents.

Land and housing that have lost their economic value in the framework of the capitalist value generation process can be converted into spaces of possibilities if the perspective on the 'ruins' changes and people with outside perspectives reconvert places of decay into places of creativity. The global conversion of post-industrial spaces into creative hubs shows that ruins can be converted in the framework of capitalism and gentrification. But a conversion of decayed spaces in the context of less developed regions can also be achieved if creativity is stimulated by the prospect of land- or house (co-) ownership and the possibility to achieve relative economic freedom and spaces of possibility. Retopian projects can thus link creative people to spaces with a potential for development, agency, and participation.

The Detroit case shows that ruins can be partially activated, albeit in a process of neoliberal gentrification. The ruins were revitalized because of their aesthetic validation by some artistic-minded groups who saw beauty in decay. Of course, this revival was limited to groups of mostly middle-class bourgeois bohemians, and the exclusion of the poor was exacerbated by the market-driven logic of this revitalization. The question is whether aesthetic revalidation is possible in an inclusive logic which allows for a wider participation in the benefits of development.

The overall idea to give a new identity to post-industrial and/or ruined spaces was born in the 1950s with the loft. Alexandra Lange describes the process through which a two-block long street in Manhattan called the Coenties Slip was revitalized when groups of artists discovered that the departed spaces offered the ideal location for studios. Cashless painters like Barnett Newman, Robert Rauschenberg, and Jasper Johns dwelled in these spaces deemed irrelevant by the market. The idea inspired other creatives, and thus other areas were developed. The most significant change happened in a decayed area then called 'Hell's Hundred Acres' where dilapidated five-story buildings with big windows and a lot of daylight were rediscovered as artist studios. Former industrial ruins were inhabited by people who appreciated the low rental price, the space, and the new aesthetic dimension. 'Hell's Hundred Acres' thus became Soho, the name under which the district is now known. The form of habitation transformed into a desirable style, later imitated and commercially exploited. As Alexandra Lange explains:

> And thus the loft was born: as much a workshop as a residence, as much a claiming of territory as the making of a home, as much about square footage as about exposed brick, high ceilings, and cast-iron columns. Creative people looking for room to experiment, and willing to live outside the domestic norm, found homes in buildings made for mass production – and accidentally created a new style of living.[31]

The dwellings out of necessity, born in an experiment, became lifestyle icons with a profound influence on the idea of 'creative living' even in the 21st century. Developers build new 'loft-style' condos even in the suburbs. The loft-dweller of today is buying an idea of creativity and pays a high price for it. The aesthetic validation of former industrial 'ruined' districts is a worldwide phenomenon. Ruins become revamped by profit-hungry developers who offer an aura of 'artistic living' to people who spent their days in grey offices. As a result, Soho today would be unaffordable for an emerging artist. Hell's Hundred Acres was transformed into a lifestyle paradise for the affluent.

In a retopian context, the aesthetic revalidation process, in order to be socially equitable and inclusive, has to be disconnected from the capitalist value creation process. As the 'loft story' shows, the aesthetic dimension, the idea of a 'creative lifestyle', is an important pull factor for neighbourhoods in decay. Animated by images, future images, and lifestyle images, people are ready to engage with rubbles and ruins. Revitalization and revalidation processes in abandoned areas are essentially connected to the aesthetic dimension. The most important question is how to replicate the 'loft' process without the forces of the market.

Jens Beckert describes how capitalism is linked to the logic of a temporal orientation toward an open future and thereby co-opted the human ability to imagine future states for purposes of commercial exploitation. Beckert writes:

> As they seek to make profit, augment their income, or increase their social status, actors create imaginaries of economic futures, the achievement or avoidance of which motivates their decisions. The temporal disposition of economic actors toward the future, and the capability to fill this future with counterfactual economic imaginaries, is crucial to

> understanding both how capitalism diverges from the economic orders that preceded it and its overall dynamics.[32]

Beckert links fictional expectations to dynamism, albeit a dynamism that is reduced to the idea of financial gain and competitive edge. The fictional expectations of the (economic) actors are thus a decisive factor that influence the decisions about the future. Expectations thus form spaces of mental possibility. Confidence in fictional expectations leads the actors, as Beckert underlines, 'to act as if the imaginary were the "future present"'.[33] Fictional expectations undermine the rational expectations theory for they are not based on established facts. In times of increasing complexity, information overload, and growing epistemological insecurity, the decision-making process cannot be based on purely rational expectations anymore. The future, not only in business, cannot be determined by all-encompassing knowledge and thus become the domain of (speculative) expectations. And expectations are grounded on counterfactual assumptions since they are directed towards an open future. So, even in the domain of business, according to Beckert, 'fictional expectations take a narrative form, and become articulated as stories that tell how the future will look and how the economy will unfold into the future from the current state of affairs'.[34] Beckert emphasizes that:

> capitalism is an economic system in which the present is assessed principally through the lens of the future, which is itself considered using imaginaries of future states in order to anticipate as yet unrealized profit and loss.[35]

The driving force is thus a belief in the future that is formed by fictional expectations, which are grounded on counterfactual assumptions, which are based on stories. The success of Elon Musk would be unimaginable without the fictional expectations he created.

For the success of the revitalization of peripheric spaces, fictional expectations are thus essential. The aesthetic dimension plus the fictional expectations are creating a mind scape of possibility. Learning from capitalism thus means creating fictional expectations that are inspiring people. But the motivation is not profit maximization but the creation of a new livelihood.

With a shift in fictional expectations from profit maximization to social equity and well-being, Retopian imaginaries also reflect degrowth and post-growth perspectives. In the 1970s, André Gorz, in the framework of his 'écologie politique' called for 'décroissance' (later translated as degrowth). The term was largely used as a reaction to the 'limits to growth' that the Club of Rome in its eponymous 1972 report mapped out. The concept became more relevant in early 2000s. Schneider, Kallis, and Martinez-Alier define degrowth as 'equitable down-scaling of production and consumption that increases human well-being and enhances ecological conditions'.[36]

Demaria, Schneider, Sekulova, and Martinez-Alier define six sources of degrowth: ecology, anti-utilitarianism, meaning of life and well-being, bioeconomics, democracy, and justice. Degrowth is thus criticizing the current GDP growth-based development models on multiple levels. In opposition to concepts like decoupling, green growth, and sustainable development, degrowth proposes a *res communis* approach in which environmental goods are commonly cared for. On an economic level, the movement positions itself against the production and consumption models of capitalism, exaggerated faith in technological fixes for societies, and uniformization of cultures. With the prospect of automation and the redundancy of large parts of the working population, the alternatives to the utilitarian approach are sharing economies and the focus on social relationships. Serge Latouche asserts:

> Just as there is nothing worse than a work-based society in which there is no work, there is nothing worse than a growth-based society in which growth does not materialize. And that social and civilizational regression is precisely what is in store for us if we do not change direction.[37]

Martinez-Alier calls for a planned degrowth 'in order to find a prosperous way down'.[38] But downscaling in economic terms can be done in a way that enhances the well-being if the patterns of work and consumption are redefined. Bioeconomics is an attempt to slow down the material degradation and diminish the ecological footprint.

Degrowth also focusses on introducing a more direct and participatory democracy based on high levels of autonomy. In this understanding, justice consists of large-scale redistribution and the reduction of excessive incomes and wealth in order to create more social equity.

As an important step towards degrowth, Serge Latouche emphasizes the aspect of relocalization, which includes local production, local consumptions, and local communities. Latouche explains:

> Whilst ideas must be able to ignore frontiers, the movement of commodities and capital must be restricted to essentials. If we wish to build a serene degrowth society, relocalization is not just an economic issue: politics, culture and the meaning of life must rediscover their local roots. This implies that all economic, political and cultural decisions that can be made at the local level must be made at that level.[39]

Effectively, relocalization is meant as an anti-globalization (in economic terms) and an anti-free trade statement. The argumentation for free trade and (economic) globalization is the growth imperative, hence degrowth reasserts local practice as an antipode.

It is more than an interesting detail, that Latouche's 'decisions that can be made at the local level must be made at that level' knowingly or unknowingly reflects one of the founding principles of the European Union: the principle of subsidiarity. Article 5 of the Treaty of the European Union states that 'the Union shall act only if and in so far as the objectives of the proposed action cannot be sufficiently achieved by the Member States, either at central level or at regional and local level, but can rather, by reason of the scale or effects of the proposed action, be better achieved at Union level'.[40] The subsidiarity principle was thus enshrined in 1992 in the European Union treaties. In 1989, Jacques Delors, the eighth president of the European Commission underlined 'the absolute necessity to preserve our roots in the shape of nations and regions' and emphasized that 'we should never entrust to a bigger unit anything that is best done by a smaller one'.[41] Other European politicians such as Jacques Santer, Romano Prodi, José Manuel Barroso, Martin Schulz, Angela Merkel, and Jean-Claude Juncker have since reaffirmed the importance of the subsidiarity principle but without putting the principle into action. Theoretically, and in speeches of political leaders, subsidiarity matters in the European Union, but the growing tendency to concentrate more executive and legislative power in Brussels contrasts with intentions to grant more autonomy to the regions. As Jean-Pierre Danthine observes, 'Europe's deep-seated institutional design problem is tied to the inevitable trade-off between efficiency-enhancing centralisation and democracy-enhancing sovereignty'.[42] Danthine underlines that the subsidiarity principle lacks precision and content therefore risks 'being buried under the centralization culture of many European politicians and the Leviathan tendencies of Brussels'.[43]

As an alternative to centralization, Latouche underlines that 'relocalization has a central role to play in our concrete utopia'[44], which is not 'a closed microcosm, but a knot in a network of virtuous and interdependent transversal relations, with a view to experimenter practices that can strengthen democracy'.[45] The economic foundations are self-sufficiency in food and financial self-sufficiency and a focus on local production, distribution, and consumption. There is a perceptible nostalgia when Latouche quotes Michael Singleton's description of degrowth:

> an organic 'reweaving' of the local (allowing people to spend more time together, as they did until the 1960s, thanks to, amongst other things, village schools, 'family' firms, local shops, and cinemas rather than spending their days shuttling between schools, industrial zones, and out-of-town supermarkets).[46]

The question is whether this nostalgic imaginary can be achieved in a time when the economic foundations are completely different from the era of post-war growth in Western Europe. But for retopian imaginaries, the question is how to achieve the 'reweaving' of the local in a 21st-century context.

Since the Covid 19 pandemic, there has been a noticeable shift from degrowth to post-growth approaches. Post-growth is based on the same critique of GDP-oriented wealth creation, like degrowth. It has the same focus on sharing, collaboration, and social justice, but is more concentrated on both open experimentation and the development of existing concepts, technologies, and ideas. Post-growth has also a strong focus on urban planning and the usage of the increasingly limited resource of land. As Christian Lamker points out, post-growth 'means acting in circular ways with available resources (including land) through a multiplicity of roles, with the aim to enhance the quality of life for a diversity of people'.[47] A lot of initiatives can be understood as both degrowth and post-growth for example buy local, seasonal eating, passive solar design, voluntary simplicity, Gross National Happiness, seed swapping, car pooling, asset-based community development, collaborative consumption, local exchange trading systems, community-supported agriculture or Transition towns.

While all of these initiatives have an often tremendously positive impact on both humans and nature, the question of the land, its ownership status, its potential for use, and its importance in the process of relocalized community creation has not been addressed.

The importance of land and its magnetic attraction and the dynamism it produces cannot only be illustrated through the example of the mass migration movements to the United States in the 18th and 19th centuries but also by a specific event in American history. On 22 April 1889, at noon in Guthrie Oklahoma, a unique event took place: the biggest race for land in human history. Fifty thousand people were standing at the starting line, waiting for the signal to rush out in order to gain a piece of the 8100 km^2 of unassigned land. The most spectacular transformation happened in Guthrie itself, according to Simon Winchester, who said the 'speed of the city's transformation beggars belief'.[48] In the morning of that day, Guthrie had zero inhabitants; in the evening it boasted more than 15,000. As the Oklahoma land rush demonstrated, the prospect of land can have electrifying power. Of course, it is also an example of exclusion and land robbery: the land was Native American, and the Native Americans were excluded from the race on their own land. In addition, the logic of land distribution was not only racist but profoundly capitalist: competition, focus on private property, and 'the winner takes all'.

But in the 21st-century context, the electrifying power of land is a potential force even with a post-growth/degrowth perspective: while today there are no unassigned lands, there is a lot of available land. And there are a lot of cities and villages that have already an established infrastructure (unlike Guthrie in 1889). Where are the 50,000 land-hungry people waiting on the starting line? And how can the new 'land rush' be organized with fictional expectations that are not neoliberal, market-fundamentalist, and profit-oriented but inclusive, participatory, and with a focus on the commons?

So, what could be a scenario that might activate people to engage with the adventure of the land in the periphery? The Covid 19 pandemic has already led to a flight from urban centers due to the lack of space during the lockdown. In addition, the new options for smart working have enhanced the relocalization options. The question is how to make the project of moving into the periphery more attractive, how to create viable options? Is there a possibility to create a 'loft effect' without the consequences of gentrification? What are the options to revitalize the periphery in the context of new work arrangements that are not smart working but possibly 'post-working'? And how can these arrangements be made in a context of post-growth?

Michel Godot and Fabrice Roubelat describe the basic mind-set needed to anticipate change:

> In the case of the accelerating pace of change, the uncertainties of the future, and the increasing complexity of phenomena and interactions, an antifatalistic, pre-active (anticipating changes) and pro-active (provoking changes) attitude is essential.[49]

In the context of the dangers of environmental degradation and the crisis triangle – depopulation of the periphery, new 'useless class', obsolete social welfare systems – a pre-active approach means attracting groups with projects and retopian imaginaries before the decomposition of the labor market happens and before the periphery falls into hopeless decay. The proactive attitude would be to develop alternatives to the market fundamentalist approach, to find different ways of engaging with the life-world and to show that change is possible, not only in policy programmes, political discourses, and activism but through a functioning experiment which can serve as an inspiration to other people. Of course, the story of an open experiment cannot be scripted. But in terms of scenario planning, the spaces have to be created in order to make them possible.

Wendell Bell defined the future in three categories, 'what can or could be (the possible), what is likely to be (the probable), and what ought to be (the preferable)'.[50] The retopian experiments in the periphery are in the realm of the possible and, from a normative point of view, if the premise is to find alternatives to the social, economic, environmental, and political destruction of neoliberalism, they certainly are preferable. Whether they are likely to be depends on a collaboration between interested groups of new periphery dwellers, local administrations in the periphery, and public sector collaboration with the nation state, in which the periphery is located, and the European Union.

As Kees van der Heijden has pointed out, scenarios broaden the perception and provide mental models to see and perceive the outside world. Heijden explains that a scenario 'provides a language through which the resulting issues can be discussed in the organization, new theories of action can be jointly developed and shared, and alignment of mental models can be achieved, necessary for institutional action'.[51] Even if retopian projects cannot be strategically planned, like scenarios for businesses or other institutions, the retopian scenario can provide a language and a theory of action and a mental model to enable citizen participation and institutional action. Van der Heijden proposes to develop a story line that provides the contextual framework and maps out the progression of cause and effect over time.

Michel Godet, describing the scenario in the context of the strategic prospective approach, highlights the five functions of scenarios: 'stimulate the imagination, reduce inconsistencies, create a common language, structure collective thought, and enable appropriation by decision makers'.[52] There is a common ground between Godet and Van der Heijden. The stimulation of the imagination corresponds to Van der Heijden's mental models, the reduction of inconsistencies to the progression of cause and effect over time. Both, Godet and Van der Heijden highlight the collective dimension and the communication with institutional decision makers.

But, retopian scenarios strongly diverge from strategic, prospective approaches or strategic planning in the sense that their speculative and normative dimension is stronger. Thomas More, in that understanding could be seen as a scenario writer. His book *Utopia*, despite the ambiguity between 'good place' and 'no place' and despite the discussion whether it was meant to be a purely fictional account or not, was a political scenario. Unlike *Utopia*, retopian scenarios are located in the here and now, not on a fictional island. Unlike Hythloday, the protagonists of the 21st-century journey cannot travel to an 'ideal commonwealth' but have to construct it themselves in the ruined periphery.

From a storytelling point of view, the retopian scenario is thus not a utopian version of what Christopher Booker calls a 'voyage and return' plot, in which the protagonist experiences a strange abnormal world and then returns home but a 'quest' plot. According to Booker, the quest 'centres on the pull of the hero towards some distant, all important goal'.[53] Booker maintains that each quest begins with a call which is the reason why the hero and his companions set out on their journey: 'Life in some "City of Destruction" has become oppressive and intolerable, and the hero recognises that he can only rectify matters by making a long, difficult journey'.[54] In addition, 'times are out of joint',[55] and something 'has gone seriously and terrifyingly wrong'.[56] So the heroes have to aim for a distant, life-renewing goal in order to fix the situation.

Times are decidedly out of joint, and the contemporary 'heroes' do not only find the city of destruction far away but their 'home lives' are also becoming increasingly complex, with the risk that the heroes lose the one thing that is essential for a good (life-) story: meaning. Be it through automation or, paradoxically, burnout due to overworking, more and more 'heroes' are feeling useless. The 'city of destruction', so the abandoned peripheries, are providing the potential to provide the distant, life-renewing goal but the difficult journey ahead is how to disentangle from the 'old life'. The solution: the hero needs companions. Booker underlines that 'a distinctive mark of the Quest is the extent to which, more than in any other kind of story, the hero is not alone in his adventures'.[57] Of course, in order to find the companions, the hero has to overcome his old patterns of individualist thinking and open up to engage with groups of people, networks, and create a common project, a sort of map, for the journey.

Then, as in every quest story, there are the helpers. In the quest plot, according to Booker, in addition 'to all the negative figures the hero and his companions meet on the journey, they also, as we have seen, encounter some very different figures: the "helpers" who give them positive assistance'.[58] The helpers in the retopian journey should be the institutions who, enabled through data taxes, robot taxes, financial transaction taxes, and the re-routing of some social-welfare money, are helping to finance the journey. The European Union, who also as an acknowledgement of the subsidiarity idea and as regional development in the periphery, co-finances the retopian projects. The national states, who set up funds to give away free land and housing in structurally underdeveloped regions to the hero and his companions. In addition, the legislative framework has to be modified to allow more regional autonomy. The regions support the projects with local know-how, a simplification of planning procedures, and the reduction of other administrative obstacles. All of these objectives are in correspondence with European Union goals which is fixed in a common decision by the Council and the European Parliament in 2013 the no net land take (NNLT) objective. The target of the NNLT policy is to halt the net-land take completely by 2050. It is impossible to reach this goal without reactivating disused land, revitalizing peripheries, and activating the 'leftover' spaces of capitalism.

The quest plot thus continues. After the call, the journey starts, followed by arrival and frustration. To reconstruct 'ruined' spaces is a task which requires heroic efforts, and many dreams and illusions will fade away. People will leave, give up, return to their home destinations. But the heroes will continue. They will be challenged by the final ordeal – often three tests and potentially a great battle – before they reach their goal. The tests are local specific: it could be attacks from locals who refuse to welcome the newcomers, environmental catastrophes, catastrophic lack of infrastructure, group incoherence, lack of skills to deal with the local tasks, or reconstruction burnout. But, once the tests are passed and the battle is won, the heroes and their companions have reached their (life-renewing) goals: a life on completely new foundations, inner change, and a new form of freedom. As Booker sums up, the quest plot begins:

> on a note of constriction, followed, when the hero or heroine respond to the 'Call', by a phase of expansion, as spirits and hopes are lifted. This leads eventually to a more serious constriction, leading to a phase when the hero or heroine are gradually being brought to a state of readiness for the final decisive confrontation with the dark forces which have so long oppressed them. When this arrives, providing the climax to the story, constriction reaches its height. Then comes the reversal, the triumphant liberation which paves the way to the happy ending.[59]

Of course, in life, unlike in fiction, happy endings are often the beginning of a new quest. But the quest plot is essentially about a group of heroes with a call on a difficult journey to a life-renewing goal. The retopian story is thus, in its emotional drive, a story of hope. For Jonah Sachs, it is the stories of hope that have the potential to shape culture. Sachs writes:

> So many of the stories that have really stuck, that have shaped culture, are about one thing: people reaching for their highest potential and struggling to create a better world. If the test of time is our judge, stories with this formula have a near-monopoly on greatness.[60]

In Sachs' definition, a great story is providing an explanation of the problematic of the world and a meaning which gives an orientation on what to do about the problem. The story encompasses explanation and meaning through compelling heroes and villains, a well-defined conflict, high stakes, and a journey towards empowerment.

The setting of the story in retopia is clear: the ruined peripheries. The heroes are groups 'social innovators' who are going on a land rush with a project of society. They share common values, share fictional expectations, and imaginaries of a better society. Their motivation to go on that journey can be manifold: curiosity, unfulfilled desires, need for a new challenge, feelings of uselessness, creative adventurism... The goal is to create a space, be it a village, a small island, or a district of a city, in which they can find (their own definition of) freedom in a group. The villains are the people who, in a drive for profit-maximization, are destroying the social, political, ecological, and ecological foundations of the planet.

The conflict is to prove that a different life is possible to make, against all obstacles, an alternative to TINA work. Not through policy declarations or capitalism critique without consequence but through shared lived experience. The stakes are high: without a viable alternative, without projects that demonstrate on a local scale that different society arrangements can function, the monopoly of neoliberal orthodoxy in all its facets (including the state-managed more authoritarian version of neoliberalism) will degrade the possibilities of livelihood even further. As the quest plot shows, the obstacles are high, since outer and inner struggles can destroy groups. The journey towards empowerment is filled with challenges, and unlike in Thomas More's book *Utopia*, the ideal commonwealth is not just appearing out of nowhere but is an ongoing process with ordeals on the way.

The motivation and the explanation of the quest plot: neoliberalism and profit maximization don't work. It destroys the fundaments of human well-being. In addition, it leads to a monopolization of the mind space, through an elimination of alternative ideas in what Graeber called the 'war on imagination'. The meaning of the retopian story is thus: prove that the mind space can be opened again by finding groups, developing projects, and going out there into the periphery to build new worlds. Experiment with society set-ups, connect to people, find new ideas – not for economic gain but for the sake of improving the life of your group. The 'triumphant liberation' cannot be found through individualism but through some form of collaboration. Friedman's claim that neoliberalism is 'whatever they value' can be also turned into the opposite of neoliberalism when it is understood as a group challenge. Friedman, the most prominent figure of the Chicago School of Economics, had a colleague in the department of Sociology, who retired 13 years before he took his first teaching job: Robert E. Park. They thus never met in the corridors of the university, but if they had, Park would have probably asked from Friedman:

> Does the community give its people security, present and future? Does it make life interesting or is life, in spite of all its opportunities, dull? Does the community offer every individual, somewhere and in some group, a status, a place in which he feels that he functions and in which he can have a certain honest pride?[61]

Unfortunately, Milton Friedman, neither as a young lecturer nor as the old Nobel prize-decorated father of neoliberalism, would have had no answer to these questions, since they are not relevant in the Friedmanian universe. But any society today would have to attempt to find a good answer to that question.

The core of Ruth Levitas' Imaginary Reconstitution of Society is the idea that (utopian) ideas can have a catalyst function, and new society models can be developed. New ideas can lead to a new ontological mode when they are implemented. Society can be reconstructed, first on an imaginary level, but then also on an ontological level. This is the optimistic message of the Imaginary Reconstitution of Society. Shared imaginaries can lead to new societies.

The good news is that, for the retopian, 'the quest plot', some preconditions for a successful hero's journey are already there. Most importantly, the spaces of possibilities are not in the imaginary realm, not in a non-place. They can be located on a (European) map. What is yet missing are the ideas, about how to explore these spaces, and the social innovators with a willingness to embark on the journey.

Notes

1 Geoff Mulgan, 'Innovation in 360 Degrees: Promoting Social Innovation in South Australia', *Adelaide Thinkers in Residence*, June (2018), pp. 9–10.
2 Ibid., p. 14.
3 Peter De Smedt, Kristian Borch, and Ted Fuller, 'Future Scenarios to Inspire Innovation', *Technological Forecasting and Social Change*, 80(3) (2013), p. 5.
4 Michel Godet, *From Anticipation to Action: A Handbook of Strategic Prospective* (Paris: Dunod, 1991), p. 2.
5 Ernst Bloch, *The Principle of Hope* (reprint ed. 1995), trans. by N. Plaice, S. Plaice, and P. Knight (Cambridge, MA: MIT Press, Vol. 3:3, 1986), p. 198.
6 Bloch, p. 186.

7 David Harvey, 'Globalization and the 'Spatial Fix'', *Geographische Revue* (2/2001), p. 24.
8 Lewis Mumford, *The Story of Utopias* (New York: Boni and Liveright, 2008, original work published 1922), pp. 184–186.
9 Zygmunt Bauman, *Retrotopia* (Cambridge: Polity, 2017), pp. 166–167.
10 Barbara Goodwin and Keith Taylor, *The Politics of Utopia: A Study in Theory and Practice* (London: Hutchinson & Co, 1981), p. 109.
11 Ibid., p. 213.
12 Molly Nesbit, Hans-Urlich Obrist, and Rikrit Tiravanija, 'Meeting Immanuel Wallerstein', in *Utopias: Documents of Contemporary Art*, ed. by R. Noble (London: Whitechapel Gallery and Cambridge, MA: MIT Press, 2003), pp. 189–191 (p. 190).
13 Bertrand De Jouvenel, 'Futuribles' (1965) <http://www.rand.org/content/dam/rand/pubs/papers/2008/P3045.pdf> [accessed 10 July 2021]
14 Wolfgang Pircher, 'On the Construction of Worlds: Technology and Economy in European Utopias', in *Thinking Utopia*, ed. by J. Rüsen, M. Fehr, and T.W. Rieger (New York: Berghahn Books, 2005), pp. 67–86 (p. 81).
15 Mumford, p. 166.
16 Ruth Levitas, 'The Imaginary Reconstitution of Society: Or Why Sociologists and Others Should Take Utopia More Seriously', Inaugural Lecture, University of Bristol (24 10 2005) <http://www.bristol.ac.uk/depts/sociology/staff/pubs/levitasinaugural> [accessed 12 February 2019]
17 Ruth Levitas, *The Concept of Utopia* (Witney: Peter Lang, 2011), p. 6.
18 Ibid., p. 40.
19 Mumford, p. 12.
20 Max Haiven and Alex Khasnabish, *The Radical Imagination: Social Movement Research in the Age of Austerity* (Halifax: Fernwood Publishing and London: Zed Books, 2014), p. 210.
21 Haiven and Khasnabish, p. 23.
22 Ibid., p. 3.
23 Ibid.
24 Fred Polak, *The Image of the Future* (Amsterdam: Elsevier, 1973), p. 302.
25 De Smedt, Borch, and Fuller, pp. 432–443 (p. 438).
26 Ibid., p. 439.
27 Frank Moulaert and Pieter Van den Broeck, 'Social Innovation and Territorial Development', in *Atlas of Social Innovation: New Practices for a Better Future* ed. by SI Drive (Brussels: SI Drive, 2017), p. 29.
28 Ibid., p. 25.
29 Umberto Pisano, Lisa Lange, and Gerald Berger, 'Social Innovation in Europe. An Overview of the Concept of Social Innovation in the Context of European Initiatives and Practices', *ESDN Quarterly Report* n. 36 (2015), pp. 6–7.
30 Mulgan, p. 8.
31 Alexandra Lange, 'How the New York Loft Reclaimed Industrial Grit as Urban Luxury', *Bloomberg* (2022) <https://www.bloomberg.com/news/features/2022-06-22/a-design-history-of-the-new-york-loft> [accessed 15 August 2022]
32 Jens Beckert, *Imagined Futures: Fictional Expectations and Capitalist Dynamics* (Boston: Harvard University Press, 2016), p. 2.
33 Ibid., p. 10.
34 Ibid.
35 Ibid., p. 22.
36 Francois Schneider, Giorgos Kallis, and Joan Martinez-Alier, 'Crisis or Opportunity? Economic Degrowth for Social Equity and Ecological Sustainability', *Journal of Cleaner Production*, 18(6) (2010), pp. 511–518.
37 Serge Latouche, *Farewell to Growth* (Cambridge: Polity, 2009), p. 8.

38 Joan Martinez-Alier, Unai Pascual, Frank-Dominque Vivien, and Edwin Zaccai, 'Sustainable De-Growth: Mapping the Context, Criticisms and Future Prospects of an Emergent Paradigm', *Ecological Economics* 69(9) (2010), pp. 1741–1747 (p. 1745).
39 Latouche, pp. 37–38.
40 'Consolidated Version of the Treaty on European Union, Title I – Common Provisions, Article 5 (ex Article 5 TEC)', EUR-Lex <https://eur-lex.europa.eu/legal-content/EN/TXT/?uri=celex%3A12016M005> [accessed 10 July 2022]
41 Rudolf Rayle, 'What Happened to EU Subsidiarity?', *World Economic Forum* (11 December 2015) <https://www.weforum.org/agenda/2015/12/what-happened-to-eu-subsidiarity/> [accessed 10 July 2022]
42 Jean-Pierre Danthine, 'Subsidiarity: The Forgotten Concept at the Core of Europe's Existential Crisis', Centre for Economic Policy Research (12 April 2017) <https://cepr.org/voxeu/columns/subsidiarity-forgotten-concept-core-europes-existential-crisis> [accessed 20 July 2022]
43 Ibid.
44 Latouche, p. 44.
45 Ibid., p. 47.
46 Ibid., p. 47.
47 Christian Wilhelm Lamker, *Becoming a Post-Growth Planner* (Rooilijn: University of Groningen, 2021) <https://www.rooilijn.nl/artikelen/becoming-a-post-growth-planner/> [accessed 22 July 2022]
48 Simon Winchester, *Land* (London: William Collins, 2021), p. 156.
49 Michel Godet and Fabrice Roubelat, 'Creating the Future: The Use and Misuse of Scenarios', *Long Range Planning*, 29 (1996), pp. 164–171 (p. 164).
50 Wendell Bell, *Foundations of Future Studies* (London: Transaction Publishers, 1997), p. 73.
51 Kees Van der Heijden, *Scenarios: The Art of Strategic Conversation* (New York: John Wiley & Sons, 1996), p. 113.
52 Michel Godet, 'The Art of Scenarios and Strategic Planning: Tools and Pitfalls', *Technological Forecasting and Social Change*, 65(1) (2000), pp. 3–22 (p. 8).
53 Christopher Booker, *Seven Basic Plots: Why We Tell Stories* (London: Continuum, 2004), p. 83.
54 Ibid., p. 83.
55 Ibid., p. 70.
56 Ibid.
57 Ibid., p. 71.
58 Booker, p. 77.
59 Ibid., p. 85.
60 Jonah Sachs, *Winning the Story Wars* (Boston: Harvard Business Review Press, 2012), p. 3.
61 Robert E. Park, 'The Significance of Social Research in Social Service', *Journal of Applied Sociology*, 8 (1924), pp. 236–237 (p. 237).

Bibliography

Bauman, Zygmunt, *Retrotopia* (Cambridge: Polity, 2017)

Beckert, Jens, *Imagined Futures: Fictional Expectations and Capitalist Dynamics* (Boston: Harvard University Press, 2016)

Bell, Wendell, *Foundations of Future Studies* (London: Transaction Publishers, 1997)

Bloch, Ernst, *The Principle of Hope* (reprint ed. 1995), trans. by Plaice, N., Plaice, S., and Knight, P. (Cambridge, MA: MIT Press, Vol. 3:3, 1986)

Booker, Christopher, *Seven Basic Plots: Why We Tell Stories* (London: Continuum, 2004)

'Consolidated Version of the Treaty on European Union, Title I – Common Provisions, Article 5 (ex Article 5 TEC)', EUR-Lex <https://eur-lex.europa.eu/legal-content/EN/TXT/?uri=celex%3A12016M005> [accessed 10 July 2022]

Danthine, Jean-Pierre, 'Subsidiarity: The Forgotten Concept at the Core of Europe's Existential Crisis', Centre for Economic Policy Research (12 April 2017) <https://cepr.org/voxeu/columns/subsidiarity-forgotten-concept-core-europes-existential-crisis> [accessed 20 July 2022]

De Jouvenel, Bertrand, 'Futuribles' (1965) <http://www.rand.org/content/dam/rand/pubs/papers/2008/P3045.pdf> [accessed 10 July 2022]

De Smedt, Peter, Borch, Kristian and Fuller, Ted, 'Future Scenarios to Inspire Innovation', *Technological Forecasting and Social Change*, 80(3) (2013), pp. 432–443

Godet, Michel, *From Anticipation to Action: A Handbook of Strategic Prospective* (Paris: Dunod, 1991)

Godet, Michel and Roubelat, Fabrice, 'Creating the Future: The Use and Misuse of Scenarios', *Long Range Planning*, 29 (1996), pp. 164–171

Godet, Michel, 'The Art of Scenarios and Strategic Planning: Tools and Pitfalls', *Technological Forecasting and Social Change*, 65(1) September (2000), pp. 3–22

Goodwin, Barbara and Taylor, Keith, *The Politics of Utopia: A Study in Theory and Practice* (London: Hutchinson & Co, 1982)

Haiven, Max and Khasnabish, Alex, *The Radical Imagination: Social Movement Research in the Age of Austerity* (Halifax: Fernwood Publishing and London: Zed Books, 2014)

Harvey, David, 'Globalization and the 'Spatial Fix'', *Geographische Revue* (2001)

Lamker, Christian Wilhelm, *Becoming a Post-Growth Planner* (Rooilijn: University of Groningen, 2021) <https://www.rooilijn.nl/artikelen/becoming-a-post-growth-planner/> [accessed 22 July 2022]

Lange, Alexandra, 'How the New York Loft Reclaimed Industrial Grit as Urban Luxury', *Bloomberg* (2022) <https://www.bloomberg.com/news/features/2022-06-22/a-design-history-of-the-new-york-loft> [accessed 15 August 2022]

Latouche, Serge, *Farewell to Growth* (Cambridge: Polity, 2009)

Levitas, Ruth, 'The Imaginary Reconstitution of Society: Or Why Sociologists and Others Should Take Utopia More Seriously', Inaugural Lecture, University of Bristol (2005) <http://www.bristol.ac.uk/depts/sociology/staff/pubs/levitasinaugural> [accessed 12 February 2019]

Levitas, Ruth, *The Concept of Utopia* (Witney: Peter Lang, 2011)

Martinez-Alier, Joan, Pascual, Unai, Vivien, Frank-Dominque and Zaccai, Edwin, 'Sustainable De-Growth: Mapping the Context, Criticisms and Future Prospects of an Emergent Paradigm, *Ecological Economics* 69(9) (July 2010), pp. 1741–1747

Moulaert, Frank and Van den Broeck, Pieter, 'Social Innovation and Territorial Development', in *Atlas of Social Innovation: New Practices for a better Future* ed. by Drive, SI. (Brussels: SI Drive, 2017)

Mulgan, Geoff, 'Innovation in 360 Degrees: Promoting Social Innovation in South Australia', *Adelaide Thinkers in Residence*, June (2018)

Mumford, Lewis, *The Story of Utopias* (New York: Boni and Liveright, 2008, original work published 1922)

Nesbit, Molly, Obrist, Hans-Urlich and Tiravanija, Rikrit, 'Meeting Immanuel Wallerstein', in *Utopias: Documents of Contemporary Art*, ed. by Noble, R. (London: Whitechapel Gallery and Cambridge, MA: MIT Press, 2003)

Park, Robert E., 'The Significance of Social Research in Social Service', *Journal of Applied Sociology* 8 (1924), pp. 263–267

Pircher, Wolfgang, 'On the Construction of Worlds: Technology and Economy in European Utopias', in *Thinking Utopia*, ed. by Rüsen, J., Fehr, M., and Rieger, T.W. (New York: Berghahn Books, 2005)

Pisano, Umberto, Lange, Lisa and Berger, Gerald, 'Social Innovation in Europe. An Overview of the Concept of Social Innovation in the Context of European Initiatives and Practices', ESDN Quarterly Report n. 36 (2015)

Polak, Fred, *The Image of the Future* (Amsterdam: Elsevier, 1973)

Rayle, Rudolf, 'What Happened to EU subsidiarity?', *World Economic Forum* (11 December 2015) <https://www.weforum.org/agenda/2015/12/what-happened-to-eu-subsidiarity/> [accessed 10 July 2022]

Sachs, Jonah, *Winning the Story Wars* (Boston: Harvard Business Review Press, 2012)

Schneider, Francois, Kallis, Giorgos and Martinez-Alier, Joan, 'Crisis or Opportunity? Economic Degrowth for Social Equity and Ecological Sustainability', *Journal of Cleaner Production*, 18(6) (February 2010), pp. 511–518

Van der Heijden, Kees, *Scenarios: The Art of Strategic Conversation* (New York: John Wiley & Sons, 1996)

Winchester, Simon, *Land* (London: William Collins, 2021)

6 The Practice of Possibility

The Sicilian Dream

In 2008, in the small Sicilian town of Salemi, two controversial Italian men are having dinner together that will spark an idea with a huge impact not only on Sicily but on the whole of Italy. One of the men, Vittorio Sgarbi is, at the time, mayor of Salemi, but he is not a typical provincial mayor. Sgarbi is an art historian, turned TV celebrity, turned journalist, turned Member of Parliament, turned mayor of a 10,000 inhabitant small town in the Sicilian periphery. The man he is having dinner with: Oliviero Toscani, notorious art director of the Benetton group, responsible for controversial campaigns, for instance using a dying HIV patient as a campaign motive. Both men are thus familiar with the unconventional. Sgarbi and Toscani discuss what can be done about the decaying Sicilian countryside. Young Sicilians are obliged to leave the Island because there are not enough job opportunities, investments are lacking, Sicily is increasingly becoming a place of old people, and houses and infrastructures are in a state of decay. During the evening, they come up with a simple and efficient idea: sell decaying houses for less than the price of an espresso. The idea of the 1 Euro houses was born. And Sgarbi set out to promote it to the global media. The story he offered is too good to stay unnoticed: get property on a beautiful island in Italy for a lower price than the cup of coffee you are drinking at the moment. The CNN headline "Living the Dream"[1] sees the development of a "Little America"[2] in Italy and predicts a "property stampede"[3] in Sicily. All the major media outlets in the world write articles or air programmes on the 1 Euro villages, from the BBC ("a dream to live in Italy"[4]) and the Financial Times ("dream scheme"[5]) to Fox News ("offering home buyers the chance to build their dream villa"[6]). The mayor of Salemi clearly managed to stimulate the dreams of the journalists and their audiences.

Salemi, which also has the title of 'first capital' (in 1860, Giuseppe Garibaldi made the town the capital of "Italy" for one day before the country was united, and Turin became the first capital). Once again, Salemi became the capital, this time of 1 Euro houses (at least in terms of media attention). But in 2014, the actual first sales of 1 Euro houses were made in

DOI: 10.4324/9780429288180-6

Mussomeli, another 11,000-inhabitant town in the heartland of Sicily. Ever since, according to Mussomeli's mayor Giuseppe Catania, about '250 deeds of sale to mainly foreign citizens of 18 different nationalities have been concluded'.[7] The media attention also boosted the number of visitors to the city from 380 per year to 5000. The forgotten hinterland of Sicily thus received a boost of attention. The local municipality also works on the improvement of the local infrastructure. One of the sources to tap into are Americans with Sicilian ancestors. In a return migration loop, the grandchildren of, for example, Argentinian emigrants, return to work in Mussomeli. A collaboration with the University of Rosario in Argentina was established and now Argentinian doctors work and live in the middle of Sicily.

The 1 Euro house idea spread all over Italy (and later to Spain, France, and Portugal), with houses being sold now from Liguria to Sardinia, from Lombardy to Calabria. In Sicily alone, there are currently 29 villages and small towns which are putting 1 Euro houses on sale. Most of the houses are in a derelict state and require renovation costs of tens of thousands of Euros. So, the 'house for the price of an espresso' is just a very efficient promotion image.

In 2017, Sambuca di Sicilia, a small town of 5000 inhabitants, started to put abandoned houses for sale. Formerly inhabited by 9000 people, Sambuca experienced a dramatic decline in population and opted for the 1 Euro strategy. In terms of house sales, the success was immediate: more than 60 properties were sold. As the mayor, Giuseppe Cacioppo, told The Guardian, 'between private homes and houses owned by the municipality, we have received investments totalling €1m'.[8] Cacioppo reports that within a few weeks after the announcement, the city received almost 100,000 e-mails from people expressing their interest. The idea of living a different life in the Sicilian periphery clearly resonates with people from all over the world. As a social imaginary, peripheric living, animated by all the stereotypes of Italian 'la dolce vita' (food, wine, beauty of the landscape, slow living) thus strikes a chord with what Taylor defined as 'the ways people imagine their social existence, how they fit together with others, how things go on between them and their fellows, the expectations that are normally met, and the deeper normative notions and images that underlie these expectations'.[9]

Can the underpopulated Sicilian villages with their high amount of disused and abandoned buildings thus be the place for retopian projects? In terms of aesthetic revalidation, it certainly seems to attract the fantasy of many people. While the 'land rush' happened mainly in the mailboxes of the Sicilian mayors, the 1 Euro initiative certainly contributes to the revitalization of the periphery. But the very market-driven approach of most of the initiatives creates fictional expectations that are in the realm of privatized hope. Is the Sicilian 'quest plot' ultimately turning into a mere escapist experience as opposed to being a 'hero journey', which leads to change?

Currently the Sicilian 'quest plot' has been discovered by the entertainment industry. Lorraine Bracco, the American actress ('Goodfellas', 'The

Sopranos') stars in 'My Big Italian Adventure', an HGTV production. The programme is centred around Bracco buying a home in Sambuca and refurbishing it. The BBC has jumped on the 'Sicilian 1 Euro house' train with a programme called 'The Italian Job' in which a comedian and a former 'Britain's Got Talent' judge buys a house in Salemi and instructs their audience on how to refurbish homes. The BBC promises to the audience 'two of Britain's best-loved celebrities' in an 'escapist eight-part series'.[10] The potential of the Sicilian revitalization projects is thus reduced to an experience for flat-screen TVs and couch potatoes.

With the growing global popularity, a lot of communes enter the race for attention. Cammarata even 'lowered' the price for the houses by giving them away for free. In Sambuca, Airbnb has refurbished a house and offers free rent for a year. In Poggioreale, a town left without inhabitants after an earthquake in 1968, future residents have the possibility to live in a ghost town. Calabria came up with an 'active resident project' in which some communes offer up to $28,000 for people under 40 to relocate to remote villages in order to start businesses there. The attention economy is threatening to devour Toscani's and Sgarbi's marketing idea that was meant to improve the structural problems of the Sicilian periphery.

But the Sicilian 1 Euro house story (thus far) has been largely targeted to attract attention. In that sense, it is a case of very successful storytelling that taps into similar fictional expectations, like retopian projects, and diverts them into escapist and privatized hope channels. The lack of coordination between the municipalities, the lack of strategic planning, and the insufficient linking to European Union development programmes have not explored the full potential of revitalization. Instead, the centre stage was left to real estate development with a thinly disguised ('almost for free') market logic. Is Sicily thus another case of 'ruin porn' that turns into a paradoxical cocktail of gentrification and unsolved social problems, like Detroit? Is the potential of the periphery deserted by capitalism recaptured by a neoliberal imaginary and thus thrown back into a loop with diminishing returns? Can Sicily be considered to be a failed case in the practice of possibility? Is the retopian premise 'a group of people from all over the world settles in a (half-) abandoned village in the periphery and designs a world according to their values' impossible in the Sicilian context?

As a space, Sicily offers all the possibilities: the reaction of the world media and their audiences to Sgarbi's and Toscanini's idea proves there is a huge potential. There are people who would give up their current life in Argentina, the United States, Russia, Germany, or South Africa in order to embark on a quest in Sicily. There is an abundance of abandoned spaces with the potential to restore the existing infrastructure (unlike in the case of many intentional communities which have to start from point zero), and recently the Sicilian Regional Government installed fibre optic cables even in small villages. So, potentially, the island offers possibilities for smart working and, due to the need for reconstruction, it would offer spaces for social innovators (so that they would not need to become part of the "useless class"). In addition, Sicily also lives up to all the positive stereotypes about Italy: many sunny days, beautiful landscapes, culinary know-how, slow pace of living, and a welcoming population. The island has a long tradition of mixing cultures and, due to the migration waves out of Sicily, also a profound understanding of people who leave their countries. Due to its status as a less developed region, it is possible to tap into funds for instance from the European Union. Exploring this place with a retopian imaginary, with Bloch's dictum that 'the defeated man must try the outside world again' and a different normative set-up than the focus on escapism and privatized hope, Sicily has the potential to be a space of possibility. But in order to explore the potential, it would also have to give the option of practicing the possibility: to experiment with the space and the social arrangements and to transform Sicilian villages into laboratories for societies would mean to practice possibility.

Market Place or Making Place

Together with my Italian colleague, Alessandro Nanì, who is specialized in audience studies, we set out on a research trip to Sicily. Nanì, who has Sicilian ancestors and was equally fascinated with the way in which the '1 Euro' project attracted such a worldwide audience, is animated by the question of whether Sicily could be a case for audience participation on a new level. The fact that people are changing their complete life setup in order to move to Italy is a compelling example of audience engagement in its most committed form. We met mayors of '1 Euro' villages with under 5000 inhabitants, who have not yet been interviewed by the army of foreign journalists and foreigners who live in Sicily, for in-depth interviews.

Among the many beautiful villages in Sicily is Castiglione di Sicilia. When approaching the 'borgo' by car it can be seen from afar: towering scenic landscape, the houses lining up the hill, a fortress in the middle of the green valley. The small streets of the upper town are lined with buildings from the 16th through the 19th centuries, and the many old balconies on display allow a panoramic view of snow-covered mountains in the distance and fields full of wine grapes. The beauty of Castiglione is officially confirmed by the Italian state: it is part of the circuit of the most beautiful villages and, according to a decree from 1994, considered of 'considerable public importance'. The small palazzo-style 19th-century town hall indicates that this is more than an average 2000-inhabitant borgo. The mayor, Antonio Camarda, shows us folders filled with houses in need for renovation and unfolds a map that covers the big wooden table. It shows a depiction of the town, classified according to the inhabitance status. In this 'studio di quartiere' the red dots indicate 'isolate', so abandoned. Camarda's map of Castiglione's old town is full of red dots. The study initiated by Camarda indicates that, out of the 1900 houses in the village, 935 are uninhabited.

Like many other Sicilian villages, Castiglione di Sicilia is in the process of putting 1 Euro houses on the market. Camarda presents the advantages of the village, like someone who is accustomed to promoting his hometown:

> Many people come here and fall in love with the slow lifestyle which is only 50 minutes from the metropolitan area of Catania, in 50 minutes you are at the airport. Castiglione is not just the village, there is a golf course, there are 60 wineries, 50% of the Etna doc wine is produced in Castiglione, there is Alcantara, there are hiking paths, there is the Etna. This is a country rich in history, culture, food and wine, it is a country that has a charm even compared to many other beautiful places in Sicily.[11]

Our talk with him gets interrupted, the phone rings, and a radio station in Northern Italy is doing an impromptu interview with Camarda. Again, he lists the advantages of Castiglione, adding the proximity to Taormina and to the sea on the list. Being a 'sindaco' (mayor) in Sicily in times of 1 Euro houses means being a salesman for the village, and Camarda does his job remarkably well because he believes in his village. So far, his campaign has attracted the usual 'digital land rush', Castiglione has received more than 5000 requests via e-mail. As Camarda explains: 'Let's say that behind all this there is a great promotion, a marketing activity, but then we have to define everything, we have to make a concrete offer'. But the campaign has not officially started, and thus far there are only three foreigners who live in the village: a gay couple from South Africa, who have already bought a property, and a ceramic artist from Russia who is in the process of finding the right property in the village. Camarda's development plan for Castiglione is focussed on full-time residents but also on holiday homes, Airbnbs, and hotels. So, the tourism sector, especially in relation to wine tourism and golf tourism is seen as a potentially promising sector for development. Camarda's ideal future residents are 'people looking for adventures but also people with financial means'. The goal is to 'transform the village, to make it alive, and to create job opportunities also for young people, because if the village lives it could mean that 2000 people inhabit the historic centre'. At the moment, the majority of residents are seniors, but Camarda hopes to attract young people through job openings in the tourism sector or small entrepreneurs who open restaurants. As for the mostly foreign people who are writing the 'digital land rush' e-mails to get information about the village, Camarda explains:

> 'The people we got to meet are between 40 and 50 years old. They are people who have the possibility. There are those who want to completely change their lives.' He underlines that some residents are full-time and some want 'a home for a few months.'

Younger people, 'could be attracted only if the idea of smart working starts'. Camarda himself has a seven-year-old daughter. When asked about her future in the village, he replies:

> of course, there are fewer opportunities here. It is clear that if my daughter wants to go to the theater she has to go to Catania. Here you also have a limited amount of same age peers. You can't really choose your friends. Your friends are the ones you grow up with. But in the end these friends are also real friends. The city cannot offer you the serenity that a place like this can offer.[12]

Apart from the climate, the beauty of the landscape, and the culinary assets Castiglione thus provides a sense of community. But it is not an ideal community, as Camarda remarks, 'there is a sense of community, but one also experiences the solitude of the cities in a certain way, but in a slightly more familiar context'.[13]

Natalia, the ceramic artist from Yekaterinburg, is a woman in her early 40s. She came to Castiglione in April 2021. Her Italian is fluent, she has worked for decades in the textile trade with clothes from Italy. Mother of three daughters, two of them adult, she decided to move to Castiglione, attracted by the option of finding a cheap house in an environment she adores. Her young daughter has health problems, and the doctors advised her to live in an area close to the sea with clean air. Natalia herself has ambitious plans: with the active help of the mayor, she wants to establish a local ceramic school, an artisan network, and build a bridge between the ceramic craft of the Ural region and the Sicilian ceramic hub Caltagirone (the interview with Natalia took place a month before the start of the Russian invasion in Ukraine). She describes herself as an idealist 'open to experiment with new things.[14] In Castiglione di Sicilia, Natalia sees 'a lot of possibilities to create hope for the future'. As she explains, 'you cannot turn the world around for everyone but at least make a difference for some people'. Fascinated by the level of craft in Sicily, with the knowledge transfer between generations and with the options to collaborate, Natalia wants to show the Sicilians the 'worth of their art and the connection to other countries'. She concedes that currently 'the mentality in the Sicilian villages is a bit closed, but then again, all over the world people in villages are more closed'. Through art she aims to establish a connection, bring life to the village, and encourage people to get out of their houses: 'You have to get people out of the house. You cannot close your door if there is something interesting happening outside. There is going to be a will to return to live, to participate in the village life'. In addition to the ceramic network with Caltagirone and her own future ceramic workshop, Natalia, whose father is Jewish, is interested in the Jewish neighbourhood in Castiglione. She is in talks with the Israeli consulate in Sicily to organize events that commemorate the Jewish heritage in the town. While she acknowledges that it is not possible to turn Castiglione into a paradise, Natalia is optimistic about the future: 'I feel a vibe, a community. There are new people in the village. It is picking up'.

Itala is not part of the circuit of Italy's most beautiful villages. It is divided between Itala Marina, which is located at the Mediterranean Sea coast, south of Messina, and Itala Superiore, the old village in the mountains. The Marina consists largely of 1970s buildings in a state of disrepair. The E45 motorway cuts right through the village. On the other side, the railway line to Messina cuts off access to the beach. In Itala Marina, Sicily shows its ugly and noisy side, and even the distant view to the mountains and a few separate palm trees do not work in favour of dissipating an air of desolation. On the 3 km long way through the winding mountainous Itala Superiore a burned out car lines the street and the small river with dried out edges floats in the shadows of the mountains. Like Castiglione, the old village of Itala is in the mountains – not on top of it but in the shadowy valley side. The houses running uphill are mostly uninhabited, and many of them do not project any flavour from passed centuries but look like makeshift constructions that were made in the mid-20th century. On a cold January evening, the village is completely empty. The only striking feature: a hotel-restaurant with a big Lithuanian flag. We meet the co-owner of Le Giare, Simona, a Lithuanian woman in her late thirties. Together with her local husband Domenico, they saved the place which belonged to his father, a former mayor of the town, from bankruptcy. After a flood in 2009, the restaurant was in a state of disrepair and practically lost all of its clients. Until Domenico and Simona, both academics with PhD degrees, had an idea that brought Itala into the headlines in Lithuania. They invited the wedding planner of a renowned Lithuanian magazine together with her photographer and stylist to Itala and staged wedding photos on the mountaintop in Croce, making the village look like an idea location for weddings. From the following summer onward, many wedding parties from the Northern country took place in Itala. To the astonishment of the mostly pension-age locals, the old village with an influx of Lithuanians in fancy wedding party outfits became part of the summer picture. When the Italian media picked up on the story of the Lithuanian summer-wedding hub in the middle of the Sicilian mountains, Itala was also back on the map for the locals, and Le Giare started to flourish again.

Despite the success story Simona, a mother of two children, does not like being reduced to a wedding party organizer. In fact, according to her self-definition, she is a 'community architect'.[15] In addition to having saved the hotel-restaurant and the jobs of 10 people, Simona and her husband created an artist in residence programme, renting out holiday apartments and organizing academic seminars in Itala. Simona is also a euroguidance-ambassador, social activist, head of the association of Lithuanians in Southern Italy, and organizer of Lietuvis pas Lietuvi (Lithuanians for Lithuanians) a global network of 9000 members in 270 locations. Itala, at least in the warmer months, is thus not only hosting wedding parties but a growing number of foreigners come to visit the place as a refuge from busy city life. Journalists, academics, marketing people, activists, and artists are now discovering the pace of life in Itala.

With the help of the fibre optic cable internet, Simona wants to develop the village even further. One of her main inspirations is the former Apple employee and, now Stanford professor, Dave Evans and his ideas about Life Design. She rejects idealism but believes in the idea that everyone can create their own prototypes of themselves as a pragmatic, practical, and rational project. In Simona's words: 'I don't dream, I have goals that I reach. You decide to be someone and you become that person. A conscious life doesn't mean to expect happiness but to create it'. It took her a long time before she 'arrived' in Itala. The first years were hell because she felt isolated as a foreign woman being reduced to motherhood and the status of a housewife by the locals. It took one-and-a-half years before the village people started to talk to her. The idea of an activist and active woman with a doctoral degree did not fit into the local role models. But gradually the inhabitants of Itala started to respect the dynamic woman from the Baltics. Simona explains that her case is exemplary for many Lithuanian women in Italy. Among the 5505 Lithuanians living in the country, more than 80% are female, and most of them, despite having at least bachelor's degrees, are not working. In her association, Simona who wrote her doctoral dissertation on professional identity formation, is helping them to find a path into the Italian labour market. For the 700 Lithuanians living in Sicily, the situation is even more difficult due to the high unemployment rates on the island. When asked about the future prospects of Itala and Sicily, Simona makes a distinction between the prospects of change for the local population and the foreigners: 'From the local point, the environment will not change a lot. Sicilians are very traditional. The locals will grow their children, some will go to the North, some will come back, some will stay there'. But despite the traditionalism of its inhabitants, she believes in the development potential of the island and of Itala. She wants to establish a training center, invite more foreigners to discover Itala, and also hopes that the media attention will generate more foreign interest and new ideas: 'People will see the luxury life of simplicity that I see here and they will come. Especially those who come from busy places'. But she personally doesn't believe that the 1 Euro house projects, apart from the media attention, will bring any benefits to Itala. Currently, the village has 14 properties for sale. Simona sees her activities in Sicily and community creation, in general, as a long-term project: 'It is not enough to create communities. It is important to continue with the activities'.

Edualdo Garufi, the head of the 1 Euro house programme in Itala, confirms the digital land rush. He proudly displays 2000 printed out e-mails from people who expressed interest to buy property. 'Itala had 3000 inhabitants and now we are at 1200, the mountain area is the most depopulated, we have completely uninhabited neighborhoods in the part of Itala Superiore'[16] Garufi explains. Most of the emails come from Argentina, Uruguay, the United States, some from Germany, Finland, Australia, and Lithuania (Simona's and Domenico's wedding party activities and their participation in a Lithuanian TV show certainly helped the case). According to Garufi many of the people are between the ages of 40 and 50, often they work in jobs with remote options, such as banking, academia, and creative industries. Many of the potential buyers from the American continent have Sicilian roots, and one of the 14 houses has been already sold to a Brazilian woman with Sicilian ancestors. The local owners of the derelict properties are very positive about the project, Garufi underlines, 'for the old owners it is important to see that the property comes back to life'. One donor of a house told him that he can't wait to see the building shine again so that the sacrifices of his grandparents are not lost. The municipality of Itala has added a clause into the contract that the houses can be taken back if the owners are not renovating the property within three years. Garufi is proud of the recent achievements in terms of technology: 'We have optical fiber, even in the hilltop village we have optical fiber, all financed by European funds'. He is very optimistic that the foreigners will integrate well, learn to appreciate the slow pace of life, the zero-kilometer food, and the climate; but in order to succeed he has one important piece of advice: 'Here you live the art of getting by'.

In the North Eastern part of Sicily, in the mountains, San Piero Patti is even more remote than Itala. With its 3000 inhabitants the town is livelier than Itala, and the place is visibly inhabited. The small shops and cafés are open, groups of people chat to each other on a small local square. At first glance, no derelict houses or half-ruined houses hit the eye. The 16th-century Convent of Carmelitani Calzati, the surrounding green mountains, and the distant view of the sea and Vulcano Island provide a scenic setting. But what looks distant is in reality even more distant (more than a 1-hour car ride) due to the mountainous roads. The town is secluded on the inland. In the town hall, the assessore Marchello Cinzia explains to us that the interest of potential buyers of the five 1 Euro houses largely came from the North of Italy. The Covid pandemic has led to a reduced interest from foreigners. Covid also had an immediate effect on the interview with the mayor. Via telephone, Salvatore Vittorio Fiore, sick at home, explained to us that San Piero Patti is interested in attracting young families 'to give new life and vitality to the municipality'.[17] The potential buyers are mainly in their forties and fifties. Fiore, in the mid-term future, is targeting smart working professionals (fibre optic internet has also been installed in the town) and also wants to make the town more attractive to tourists. He wants to lobby for a new road construction to the coastline, a better connection to the airports in Catania and Palermo via the improvement of rail connection, and improve the English language skills of the locals. 'The local community is very willing to welcome people', Fiore insists, 'our community is characterized by conviviality and is open to the world'. But the mayor of San Piero Patti also understands that the development of the town is a long-term project: 'For me, the important thing is not just falling in love with an idea but also to create services and infrastructure that live up to the expectation'.

Graham is 50 years old and has a company that provides online services. Together with his Ukrainian wife, they moved to Sicily in 2019. They quickly discarded the idea of 1 Euro houses because the bureaucratic procedures were too extensive. Graham needed a fast fibre optic internet connection (which at the time only existed in big cities in Sicily), a proximity to a well-connected airport, and was looking for a house with historical charm. He bought a big house in Catania, with stucco, high ceilings, antique Majolica tiles, balconies, and a roof terrace. When Graham talks about his house, he lights up. The Brit with a rational outlook who lived in 12 countries is convinced that his digital nomad life is over, and he found himself a home. Sicily was not the initial destination for establishing the new roots. Together with his wife, they checked properties on neighboring Malta but were not only discouraged by the lack of geographical diversity and the small size of the island but also by the prices. "We have 170 m^2 with a rooftop terrace. In Malta, something like that would be probably 600,000 or 700,000 Euros. We paid 120,000 Euros. A fifth of what it would have been in Malta'.[18] Prior to his extensive house-scouting, Graham had little knowledge about Sicily. A short honeymoon in Noto. A documentary about an earthquake in the Etna region. Reports about heatwaves and legends about the Mafia. But once he analyzed the fundamentals, Graham saw Sicily as a rational choice, an 'undiscovered great place for a digital nomad life style' where there is 'so much to explore'. Graham is convinced that if 'you're rational about it: a warm place next to the Mediterranean with a major airport close by, with gigabit internet, I couldn't find anything that could compare to Catania'. The fact that Sicily is thus far so under-discovered as a residence for smart working is a mystery to Graham. Compared to the Canary Islands, Malaga, or Malta, which are overrun and overpriced in his perception, Sicily is hugely undervalued. For Graham, one of the reasons is related to a myth: 'When I told my mum about Sicily, her first reaction was: "What about the mafia?" We can laugh about that now'. The fact that residence-wise, Sicily is still unexplored in relation to its size, also leads to the fact that, according to Graham, 'there isn't really a concentration of expats – they are evenly spread across the island rather than congregating in one place'. Will Sicily thus take off and become the new destination of choice for smart workers? Graham hesitates: 'There are forces pulling in both directions. From a local's point of view, they don't have any optimism about Sicily. The economy is not good, there are no good jobs. For them it's much easier to go up North or even out of Italy to find a job'. But he strongly believes that sooner or later many people will see Sicily as a rational choice for smart working. When it comes to the role of the 1 Euro house project in revitalizing Sicily, Graham recognizes that: 'In terms of putting Sicily on the map and viral attention I think it is an absolutely amazing success. Whether it really brings people in? Maybe in some towns more than in others'. He describes the island as 'the land that time forgot' and is surprised that, in relation to the outside view of Sicily, 'it takes a long time for people to change their perception'.

Further into the mountains is Piraino, a village of 3000 inhabitants in houses that are widely dispersed around a river. In some areas, seeing your neighbour means taking a 10-minute footwalk through the mountains. Rebecca, a mid-fifties woman originally from London, rents out an apartment next to her remote house in the hills. The next convenience store or restaurant are reachable only via car or through an extensive trip for hikers with advanced skills. Before moving to Sicily five years ago, she lived in Tuscany. Comparing the North of Italy to Sicily, Rebecca explains that 'People in Tuscany are different. I have spent over 15 years in Tuscany and have two so-called friends. Here I have found many friends. It's not so superficial. It's the deeper commitment that counts here'.[19] In addition to her apartment rental, she teaches in English in a language school. She has observed that the number of foreigners moving even to remote regions in Sicily has grown over the last few years due to the good promotional work of the Sicilian authorities. Whether it will lead to change depends, in Rebecca's opinion, on the attitude of the foreigners. 'People are magical here; they have values which are hard to find elsewhere. The local community looks after you' she claims. But, if foreigners behave disrespectfully, then 'they can pack their bags', because the locals, unlike for instance in Tuscany, care deeply about respect for the community. Rebecca has integrated well; soon her girlfriend from Northern France will join her in Piraino and also buy a house.

Next to Milazzo, in a small village, Randi has bought a small unrenovated house. Three years ago, the mid-fifties Danish woman decided to move away from Copenhagen after a divorce. Her three children stayed with the father in Denmark, and all Randi wanted to do in Sicily was to forget about her previous life. The village is a refuge, she keeps her social contacts at a minimum level, lives a restrained and simple life, from alimony of her former husband and occasionally visits Copenhagen. But the deeper she dives into the slow space of her Sicilian village life, the more difficult it becomes for her to even handle the buzz of the city for very long. Randi says that her retired existence suits her well to forget about the demons of the past.

Among the many foreigners we met and talked to Natalia, Simona, Graham, Rebecca, and Randi represent some of the archetypical Sicily immigrants. Some come because they have a personal connection to Sicily or Italy, mostly partners, such as Simona and Rebecca (in her initial move to Tuscany). Some come because they are attracted by the local culture and plan artistic projects, such as Natalia. Others come because of the smart-working options, like Graham. Many are motivated by an artistic or community project, as the examples of Natalia and Simona show. These people are driven by an idea to connect to the local communities, to connect local communities to the outside world, to bring foreign know-how, to transform the communities they live in, and explore the options. Some foreigners simply come to Sicily because of what the 1 Euro house campaigns advertise: good weather, cheap prices, a slow speed of life, an escape from the lives in their home countries, such as Randi. All of them see the 1 Euro house project as a promotional tool, as a clever device to attract attention to Sicily.

But the logic of the sell-out is deeply superficial and motivated by the neoliberal imaginary. The next years will show how many foreigners find the deeper connection to Sicily once the initial magic of the 'dolce vita' has worn off. The island itself is in a critical economic situation: many young Sicilians are already now part of the 'useless class' and are thus forced to leave. As long as no meaningful mechanism can be found that connects the spaces of possibility in Sicily with the locals, and also the foreigners who wish to bring change to the island, all the initiatives risk being patches with diminishing effects. So, the change in the political imaginary in relation to the redevelopment of Sicily is crucial. The will to change is there, the international interest in the island is growing, and the 'digital land rush' shows the immense potential and attraction of the island. But the parameters of market logic and individual wish fulfillment of private hopes will make a profound change impossible.

The Retopian Imaginary and the Practice of Possibility

If practice of possibility means the exploration of the 'possible' in the sense that De Jouvenel understands the future – as a process of active anticipation – then the development of the Sicilian periphery is thus far a mere continuation of the neoliberal imaginary. The crisis triangle – the depopulation of the periphery, the development of a 'useless' class, and the crisis of the social welfare systems – is not transformed into a space of possibility. The Sicilian case shows that the potential space and the potential interest of social innovators is there, but the political choices were to opt for a market logic with attractive wholesale prices to satisfy the desires of the individualized customers and their escapist dreams.

Practice of possibility would mean allowing for social experimentation, for example, take the 935 uninhabited houses in Castiglione di Sicilia and give them away for free to a group with a specific social innovation project. In a transparent and unbureaucratic public call for offers, the space is then reallocated. The project is financed through development and social welfare funds, and the property schemes focus on new forms of commons. This space of possibility would promote social innovation by the allocation of a maximum of freedom to the 'new developers' to create spaces of participation. This 'retopian' development scheme is focussed on subsidiarity and relocalization, thus allowing for a maximum amount of autonomy. The 'retopian' development scheme would replace social welfare, in the form of social assistance or UBI, with a form of localized land redistribution that gives away peripheral spaces that have lost their market values and supports the reconstruction of communities. Potentially any community project could qualify for support: Castiglione di Sicilia could thus either become a 'health care village', a 'garden village', a collective wine-making hub, a co-operative restaurant village, a woman's village, an LGBTQ experimental sphere, a refugee care village, a tech-free village, a smart worker village, or any other imaginable form of community that aims to experiment with specific aspects of social innovation. Practice of Possibility in the context of retopian development schemes allows for localized experimentation with self-defined priorities. The priority for the allocation of real estate is thus not the market-based exchange value but a maximum variety of projects focussed on use-value and community-enhancing social, economic, and political arrangements.

Itala, with Domenico and Simona as 'community architects' could become a New Lithuania, a place where a Lithuanian (and Sicilian) community experiments with new forms of childcare, community work, and the integration of women. Free, high-quality Lithuanian–Sicilian schools and kindergartens could create an education hub, the abandoned buildings in the village being reconstructed and given for free for underpaid Lithuanian teachers who like their job but wish to work under new arrangements. The 700 Lithuanians already living in Sicily could contribute their know-how transforming an online community into a community with shared experiences. Of course, the Lithuanian aspect should not be considered an exclusion criteria but a form of open-identity creation. Anyone who is interested in child care, education, and learning a new language could become a New Lithuanian. If Lithuanian weddings already create so much media attention, a New Lithuania in the middle of the Sicilian hinterland would, if redeveloped in a retopian scheme, possibly inspire others to experiment with new forms of social arrangements. Across the Lithuanian border, in Latvia, a city like Liepaja could create their own social laboratory at the Baltic Sea coast, either experimenting with different forms of free education hubs or focussing on, e.g., the protection of maritime spaces, elder care, or tech-free living.

In Pleven in Bulgaria, a big live action role-playing (LARP) group could use the ruins of Storgosia for a permanent role play in which the participants could live either in Roman times, the Middle Ages, or whichever time period they wish to project into the space. In exchange for free living and the space to live their fantasies, they would have to contribute four hours of social work per day. In exchange, they can live the remaining 20 hours in a different time period joined by people with the same passion for the re-enactment of life in the past.

In the retopian development scheme, open local communities with specific interests can create new worlds based on local resources, develop their group-specific quest narratives, and, through their ideas, aesthetically, socially, and politically revalidate the periphery in which they are active. New stories compete with the 'same old' stories that animate the neoliberal imaginary, and the ruined periphery becomes a hub for social innovators who share the same story. The inverted spatial fix is thus stimulating multiple practice-led projects aiming at different forms of society. Pleven, Liepaja, Itala, or Castiglione di Sicilia could thus become hubs for society start-ups with a shared story, a shared location, and shared immediate experience based on the premise that action and experiment can lead to new ways of living. The ruined periphery can thus become a topoi for reconstruction, and the successful stories among these laboratories of society are real-life illustrations in the battle of the story that another life is possible. Geoff Mulgan emphasizes that two forces are interacting in social imagination:

> One is the pull of a possible future, set out in stories or accounts, acting to inspire and energize. The other is the push or movement from the fertile present-multiplying options, testing and experimenting in practice.[20]

The retopian imaginary is based both on the pull and push aspect of a possible future. Practice of possibility means that stories create local, experiment-based, possible new futures.

Ruth Levitas explains that 'Fredric Jameson has argued that the imagination of Utopia is impossible because it requires us to transcend not just where, but who, we are – and that imagining a self radically transformed holds the terror of annihilation'.[21] In retopian development schemes the self is not annihilated but individuals are able to live what is essential to them in the context of a group. It is not an imagination of utopia or a blueprint for a society but a space of possibility to open up the imagination. It appeals to a self what Will Storr calls the 'moral heroes at the centre of the unfolding plot of our lives'[22] to engage with others in a common quest narrative. David Graeber underlines that:

> Hopelessness isn't natural. It needs to be produced. If we really want to understand this situation, we have to begin by understanding that the last thirty years have seen the construction of a vast bureaucratic apparatus for the creation and maintenance of hopelessness, a kind of giant machine that is designed, first and foremost, to destroy any sense of possible alternative futures.[23]

So, the main function of the retopian imaginary, the retopian development schemes, and the creation of spaces of possibility is to recreate a sense of hope. Graeber also maintained that at 'the moment there appears to be any sense of opening, the imagination will immediately spring forth'.[24] In order to enter the quest narrative towards retopia, it is essential to suspend the disbelief, the very disbelief in possible alternative futures that has created the TINA dogma. Sicily could be one of the places to start with the quest.

Notes

1 Silvia Marchetti, 'Their Job Is Saving Lives. But Can These Argentinian Doctors Revive a Whole Italian Village?' *CNN* (29 July 2022) <https://edition.cnn.com/travel/article/mussomeli-sicily-argentinian-doctors-community/index.html > [accessed 30 August 2022]
2 Silvia Marchetti, 'One Euro Home Sell-Off has Created a 'Little America' in Italy', *CNN* (4 May 2022) <https://edition.cnn.com/travel/article/little-america-italy-sambuca-sicily/index.html> [accessed 30 August 2022]
3 Silvia Marchetti, '$1 Home Deal Triggers Property Stampede in Sambuca, Italy', *CNN* <https://edition.cnn.com/travel/article/sambuca-sicily-one-euro-home-stampede/index.html>[accessed 30 August 2022]
4 Generation Project, 'The €1 Homes of Sicily', *BBC* (19 September 2019) <https://www.bbc.com/worklife/article/20190919-the-one-euro-homes-of-sicily>[accessed 14 August 2022]
5 Zoe Dare Hall, 'A €1 Dream Scheme Sparks a Wider Homebuying Trend in Sicily', *Financial Times* (10 August 2022) <https://www.ft.com/content/fdcda477-2a09-4598-8c8b-20463170c8ac>[accessed 14 August 2022]
6 Jeanette Settembre, 'Sicilian Village Auctioning Homes for $1and Paying for Home Renovation', *Fox News* (29 January 2021) <https://www.foxnews.com/real-estate/italian-village-sicily-homes-1> [accessed 1 September 2022]
7 *Musomeli: A Success Story with 1 Euro*, 1 Euro Houses, <https://1eurohouses.com/mussomeli-a-success-story-with-1-euro-houses> [accessed 30 August 2022]
8 Lorenzo Tondo, 'Foreign Buyers Snap up Sicilian Homes at 1Euro Auction'', *The Guardian* (9 May 2019) <https://www.theguardian.com/world/2019/may/09/foreign-buyers-snap-up-sicilian-homes-at-1-auction> [accessed 30 August 2022]
9 Charles Taylor, *Secular Age* (Cambridge, MA: Harvard University Press, 2007), p. 171.
10 Alan Carr and Amanda Holden, 'Pair Up in The Italian Job', a new series for BBC One, *BBC* (20 May 2022) <https://www.bbc.com/mediacentre/2022/amanda-and-alan-the-italian-job> [accessed 25 August 2022]
11 Interview with Antonio Camarda in Castiglione di Sicilia, 17 January 2022.
12 Interview with Antonio Camarda in Castiglione di Sicilia, 17 January 2022.
13 Ibid
14 Interview with Natalia in Castiglione di Sicilia, 17 January 2022.
15 Interview with Simona in Itala, 19 January 2022.
16 Interview with Edualdo Garufi, Itala, 18 January 2022.
17 Interview with Salvatore Vittorio Fiore, San Piero Patti, 20 January 2022.
18 Interview with Graham in Catania, 18 January 2022.
19 Interview with Rebecca in Piraino, 21 January 2022.
20 Geoff Mulgan, 'The Imaginary Crisis (and How We Might Quicken Social and Public Imagination)', *UCL Demos Helsinki and Untitled* (2020), p. 22 < https://demoshelsinki.fi/julkaisut/the-imaginary-crisis-and-how-we-might-quicken-social-and-public-imagination/>
21 Ruth Levitas, *The Concept of Utopia* (Witney: Peter Lang, 2005), p. 20.
22 Will Storr, *The Science of Storytelling: Why Stories Make Us Human and How to Tell Them Better* (London: Harper Collins, 2019), p. 3.
23 David Graeber, *Revolutions in Reverse. Essays on Politics, Violence, Art and Imagination* (New York: Minor Compositions, 2011), pp. 31–32.
24 Ibid., p. 33.

Bibliography

Generation Project, 'The €1 Homes of Sicily', *BBC* (19 September 2019) <https://www.bbc.com/worklife/article/20190919-the-one-euro-homes-of-sicily>[accessed 14 August 2022]

Graeber, David, *Revolutions in Reverse. Essays on Politics, Violence, Art and Imagination* (New York: Minor Compositions, 2011)

Hall, Zoe Dare, 'A €1 Dream Scheme Sparks a Wider Homebuying Trend in Sicily', *Financial Times* (10 August 2022) <https://www.ft.com/content/fdcda477-2a09-4598-8c8b-20463170c8ac>[accessed 14 August 2022]

Levitas, Ruth, *The Concept of Utopia* (Witney: Peter Lang, 2005)

Marchetti, Silvia, 'Their Job is Saving Lives. But can These Argentinian Doctors Revive a Whole Italian Village?', *CNN* (29 July 2022) <https://edition.cnn.com/travel/article/mussomeli-sicily-argentinian-doctors-community/index.html> [accessed 30 August 2022]

Marchetti, Silvia, 'One Euro Home Sell-Off has Created a "Little America" in Italy', *CNN* (4 May 2022) <https://edition.cnn.com/travel/article/little-america-italy-sambuca-sicily/index.html> [accessed 30 August 2022]

Marchetti, Silvia, '$1 Home Deal Triggers Property Stampede in Sambuca, Italy', *CNN* (2019) <https://edition.cnn.com/travel/article/sambuca-sicily-one-euro-home-stampede/index.html> [accessed 30 August 2022]

Mulgan, Geoff, 'The Imaginary Crisis (and How We Might Quicken Social and Public Imagination)', *UCL Demos Helsinki and Untitled* (2020), p. 23 <https://demoshelsinki.fi/julkaisut/the-imaginary-crisis-and-how-we-might-quicken-social-and-public-imagination/>

Musomeli: A Success Story with 1 Euro, 1 Euro Houses, <https://1eurohouses.com/mussomeli-a-success-story-with-1-euro-houses> [accessed 30 August 2022]

Settembre, Jeanette, 'Sicilian Village Auctioning Homes for $1 and Paying for Home Renovation', *Fox News* (29 January 2021) <https://www.foxnews.com/real-estate/italian-village-sicily-homes-1> [accessed 1 September 2022]

Storr, Will, *The Science of Storytelling: Why Stories Make Us Human and How to Tell Them Better* (London: Harper Collins, 2019)

Taylor, Charles, *Secular Age* (Cambridge, MA: Harvard University Press, 2007)

Tondo, Lorenzo, 'Foreign Buyers Snap Up Sicilian Homes at 1Euro Auction', *The Guardian* (9 May 2019) <https://www.theguardian.com/world/2019/may/09/foreign-buyers-snap-up-sicilian-homes-at-1-auction> [accessed 30 August 2022]

Epilogue

Stories from an Open Future

The following fragments of what retopian development schemes in peripheral regions can look like in 2030 are speculative fiction. They represent only a small share of the possibilities and mainly serve as an encouragement to develop other stories. The start-up capital of all these social innovation projects comes from European Union structural funds, the reallocation of social welfare, and new taxes (e.g., robot taxes, data taxes, financial transaction taxes). All the projects are based on the implementation of the maximum amount of common pool resources and participatory democracy, participatory budgeting, and participatory learning. The local projects can autonomously decide on their society arrangements, but one of the prerequisites for funding is social engagement with the local communities. Thus, each retopian periphery is an independent laboratory of society.

Castiglione di Sicilia

When Chineyere and Hanna came to Castiglione di Sicilia in June 2025 they were unprepared for what would await them. Both were freelancing graphic designers for years in a creative agency in Vienna but with the arrival of new AI-generated software, the job offers came in more and more sparsely, until they stopped. Benjamin, the creative director, told them with an apologetic smile that they have to look 'elsewhere. Chineyere's and Hanna's flat-sharing community became a place of doom. Their self-isolation went to even higher levels than during the worst months of the Covid lockdowns in the early 2020s. Many of their friends still had jobs. Meeting them was a painful reminder that some people were still needed. Their disguised pity and the stale encouragements with a sympathetic attitude that bordered on the condescending were hiding an invisible wall. Gradually, the conversations were running out of steam. The days became longer and emptier. Sometimes, they visited Hanna's father in the Burgenland, and more and more often they volunteered when their neighbour needed help in the grape harvest. It was hard work, but it also helped to chase away the feeling of being useless. Then Chineyere discovered the community hub app and browsed through offers of emerging communities. When she saw the photos from Castiglione, it was an instant match. The old Italian flavor. The mountains, the volcano nearby, the idea to develop a cooperative-based wine-making project. After many online conversations and a meeting with the group which consisted of 100 people from 23 countries, they defined an action plan. The application procedure for the houses and the funding was surprisingly simple, and in January 2025 the project got approved. They called it 'Nuovo Castiglione'.

Most of the group members had no practical experience in wine-making; many of them were university graduates, and had used the preparation time to read all the relevant books on viticulture. Hanna and Chineyere were among the dozens of people that actually had harvested grapes. Pablo studied agriculture and had some courses in viticulture. But the local conditions in Sicily are very specific. The group had two months to prepare for the harvest in August. The mayor of Castiglione arranged meetings and workshops with local wine producers. Luigi, a 75-year-old local from the village, was their tutor. His 50 years of wine-making experience and his commitment to helping the newcomers were invaluable. After the long, exhausting days, Luigi's wife, Gianna, and other women of the village placed tables in the small streets and served arracini and pasta. And they were tasting wine. The village became an open-air restaurant until late at night. Coming back to their half-ruined house reminded them of another challenge ahead. Chineyere was funding the group for house restorations, and they agreed on a plan for the post-harvest season. The group decided to look out for architects and builders to join them. The first months in Castiglione were brutal despite the food, the wine, and the sense of community. Wine harvest in the Sicilian heat, arguments within the group, jealousies, ego issues, cultural differences, the question of power. After the first year, there were only 67 members left, but the waiting list was long, so it wasn't very difficult to find newcomers. Since the housing capacity was limited to 70 units, the community was restricted. They decided to gradually develop the capacity, learning from Elina Ostrom's research. The property structure they decided on was that the houses were on a permanent lease to the owner which is transferred back to the commons when the person decides to leave the community. All the income from the wine production is equally shared among the community members which also includes the Sicilian pensioners that are living in the village. Some members of the group specialize in wine-making, others on other agricultural products. The renovation process of the houses took five years, and every 'New Castiglionian' had to contribute according to their capacities. The architects did a great job in designing places that respect the local flavor while allowing everyone to find their individual solutions on the inside.

Now in 2030, life has become easier. The hours they used to spend on the renovation projects are now free for other activities. The Nuovo Castiglione wine is sold via e-shops and the reviews have been very positive. Over the years the know-how in agricultural production has also improved and the zero-kilometer food has become tasty even by Italian standards. The first

New Castiglionian babies were born, and some of the local pensioners died. But everyone takes care of the elderly, and now the open-air restaurant with free food is run by the group. The healthier pensioners take care of the children and the community takes care of the pensioners in need. Seven doctors have moved into the village. In December and January, many houses are cold because of the very basic heating systems. Some of the old women have taught knitting and, for at least two months, thick knitted pullovers are the most essential garment in the village. Hanna is taking care of the wine e-business and Chineyere, next to her urban planning activities, is also doing the PR for the village and the Nuovo Castiglione wine. The group agreed to limit tourism to wine-making workshops and social projects. Teenagers from all over the world come for internships, and other villages with similar projects can get training seminars in Castiglione. Some villages have already successfully followed the Castiglione example and the network of retopian villages is growing.

The first kindergarten group has started, and the old school building in Castiglione is soon re-opened again. They decided that the curriculum has to teach future children skills that might not be geared towards the labour market but towards life. Many of Hanna's and Chineyere's old friends who initially derided their move away from Vienna as a 'cheap Italian province escape fantasy' have lost their jobs in the last few years. When a journalist asked Chineyere what she learned from the experience in Castiglione there was a minute of silence. She sat in her chair, brushed the wrist of her woolen pullover, and scanned the Majolica floor. Then, with a smile, she replied: 'Dreaming is hard work.'

Pleven, Bulgaria

The initial idea for 'Female Rome' came from Chloé. When she realized that even after five years of temporary contracts in the history department, the University of Bordeaux would neither give her tenure track employment nor rise her 'too modest to live' salary she decided that it was time for a change. After nine years of studies of Ancient History and the odyssey of temporary contracts, Choé was tired of mid-level university bureaucrats defining her existence. She founded a group at the community hub app that her future former colleagues were laughing about: 'Female Rome'. The idea: create counter-factual Roman history through reenactment. The premise: Ancient Rome is run by women. The group quickly gained traction. Nora, a Norwegian nurse, and Katja, a German former travel consultant made redundant by booking apps, became active co-organizers, co-developers, and promoters of Chloé's idea. The group quickly gathered 280 active female members from 14 European countries. The first group meeting proved to be inspiring, many ideas for reenactment scenarios, co-living, and the implementation of 'Live Action Role Play' (LARP) in real life came into being. When the city of Pleven offered a space for a redevelopment project, the three women were skeptical. The weather conditions are not as good as in Italy, and the Roman road station in Storgosia is modest in size and historic importance for the history of the Roman Empire. But when they discovered the potential of the 10 km^2 Kaylaka Park, with its cliffs, and experienced the hospitality of the locals, they decided to develop the project with the city.

Pleven offered 300 free apartments in the centre of the city, and all of the 280 active members decided to join. The European Union offered financing for the renovation of the apartments and, most importantly, provided funds to restore the road station. Chloé's friend, Gregory, developed a plan for the redevelopment of the road station as a location for LARP, and Chloé supervised the historical accuracy. In 2027, the housing was ready and the road station retrofitted. The real-life game could start. Female Rome. The road station and the adjacent settlements provided the setting for a reenactment of Roman history. The group programmed the time machine back 2000 years. In the beginning, the costumes were inauthentic, but the simulation of a matriarchal Roman micro-society in a counterfactual mode led to agreements of temporary, limited empresses (one-month reign) with the authority to write and create the scenario of the month. Unlike temporary work agreements, temporary empresses allow for a diversity of life forms.

In the beginning, the locals were puzzled by the 300 Roman-clad women. But in a city of 100,000, they could live like a tolerated minority. The locals gradually perceived the progress around the Storgosia site and volunteered in increasing numbers to help with the restoration. The newcomers' reputation was also raised by the four hours of daily voluntary social work they provided. Nora and other nurses and doctors in the group volunteered in the hospital, and Katja organized a care group for the pensioners in need. Some volunteers taught in local schools and others joined the soup kitchen of the city. Many local children joined the LARP. For the 'Female Rome' group, the adjustment was not always easy. The local housing largely consisted of Soviet-style block houses, some areas of Pleven are not safe at night, and some of the local men derided the women as frustrated feminists. But due to the committed support of the local authorities, the group could create a parallel universe in and around the Kaylaka Park. The yearly Female Rome festival in Pleven gathered international media attention, and more and more visitors got intrigued by the real-life action role game project. During the first week in July, thousands of visitors come to Pleven. The generated income was shared between the city and the group. Attracted by the success, other LARP groups applied for projects in Pleven. 120 people converted the Pleven Panorama, a gigantic 1970s retro-futuristic building, into a site for a permanent dystopian role play set in the spy world. The biggest influx of new residents came when the city offered the Skobelev Park to a group of middle-age enthusiasts who set to create a reenactment of medieval life in the middle of 21st-century Bulgaria. More than 1500 people came to Pleven ready to trade their former existence for a 'new' life in the Middle Ages.

From 2030 onwards, the three LARP groups in Pleven co-organize the yearly summer festival which gradually transforms into a global annual event for everyone interested in reenactment. Playwrights, set designers, and directors have joined the programmes in the city, and the level of professionalism is rising simultaneously with the feeling of the authenticity of the experience. More and more people prefer the 'Pleven real player' experience even to the most sophisticated VR/AR computer games. The city becomes a synonym for the redefinition of a gaming community. Parallel to all the gaming, the ethos of social engagement with the city also has led to full acceptance of this community by the local Bulgarians.

Chloé, Katja, Nora, the pioneers of the LARP generated revival, have enlarged their group to 500 people and decided to make this the maximum to keep the group experience enjoyable. Their costumes, tools, and customs have come closer to what Chloé has researched as the everyday experience in the Roman Empire. Even if their four-hour morning shift in social assistance and their block house accommodations are firmly anchored in this century. Through the counterfactual simulation, Chloé has gained valuable insight into her research area. Every new empress brings a new facet into the 'Bulgarian Roman Empire'. Female Rome is also regularly organizing events in the Roman ruins which were part of Moesia Inferior. 200 km east of Pleven is the ancient Roman military base of Arbritus and 100 km Southeast are the town ruins of Nikopolis ad Istrum. Unfortunately, she cannot communicate her findings to the history department of the University of Bordeaux anymore. The department was closed down due to a lack of students in 2028. Chloé's applied research however is producing spillover effects in Pleven on a daily level.

Liepaja, Latvia

Things started off really well for Jasper and Anthony. Four months after they met at a party in London, their debut demo tape ends up in the BBC Sounds of 2025 list in fifth place. The concerts for their project Toco, a fusion between Arvo Pärt samples, Trinidad Dancehall Beats, poetry, and soul, are filling up with more and more people and their debut was streamed more than 1.2 million times on music platforms. The Estonian producer and the Singer from Trinidad and Tobago saw themselves on the road to the pop star 'from rags to riches' story. After Brexit, it was difficult for EU citizens to find legal employment, so Jasper was working on construction sites, ready to run whenever work inspectors showed up. Anthony's situation was even worse: he came as an undocumented immigrant with his father from Tobago when he was 11, managed to maneuver through the British school system, get an A-level in his new hometown, Harlow, and escaped to London. When he met Jasper, Anthony was not only in between low-level jobs but also between apartments, or rather between couches in apartments of different friends and acquaintances. Occasionally, he had to swap a couch for a park bench. On a musical and human level, Jasper and Anthony clicked, and Anthony's new friend and band mate offered a cupboard-size room in his mini-apartment in Croydon. But 2025 looked like a promising year for them. When 1 million streams were reached in October, Jasper and Anthony bought (their first) three bottles of (real) champagne and got drunk in the park. The hangover came months later. Their income from the million-plus streams: 7,200 Pounds. The magic that they felt when Jasper plugged in his Maschine MK3 and Anthony switched on the microphone did not even allow them to pay the bills for two months. The concert revenues also were not enough to sustain them for the whole year. Too many intermediaries were cashing in on their creativity. In a 'winner takes all' market even a fifth place condemns you to live below the poverty line. In the music industry roulette, the casino always wins.

Anthony connected to the community hub app because he was fed up with being a cash-cow for greedy music executives. The offer that caught his attention: Liepaja, Latvia. The project: a town full of independent musicians. He had to check on his map app where Liepaja is. Jasper was very negative when he told him about the Liepaja option. For him, as an Estonian, moving to Latvia, especially from London, looked like a complete downward spiral. But gradually he came around when he checked the fundamentals: free accommodation, free recording studios, excellent concert venues, and a community of musicians with over 1000 people. But what convinced Jasper most was the possibility to work with the Liepaja Symphony Orchestra and international-level choirs, like Intis and Liva. They fist-bumped the deal to move their operations to the Latvian West coast.

When they came to Liepaja in 2027, the first impression of sleepy provincialism soon wore off. In the district of Karosta, a former Russian military base founded during the tsarist era, the sheer amount of bustle of people from all over the world, the sounds coming from rehearsal rooms, the jamming sessions at the beach, and the overall energy of the place was unexpected – even for Jasper, who knew the town from a family vacation 15 years ago. The prison was transformed into a hub for recording studios, the water tower hosted the common concert pool agency which is in charge of organizing at least three concerts every day in the city. The Karosta Manege, a former tsarist era horse-riding arena, can seat 4000 people for open-air concerts in a unique atmosphere. The former Station of Homing Pigeons, a big red-brick end of the 19th-century construction, hosts apartments. Many musicians also live in refurbished block houses. The idea to unify two floors, by taking the in-between ceiling away, transformed the Soviet-era boxes into light and spacious residences. Altogether, more than 1000 musicians live in Karosta.

Every age, instrument, and musical style is represented, and the proximity of the people quickly generated a unique spirit of collaboration. Since the income from the concerts is shared equally among the residents, the spirit of competition is limited to the question of who is the most creative.

Karosta is not only a laboratory of society but also a laboratory for new music. Anthony and Jasper feel the openness and collaborate on many projects. Jasper's dream comes true: in the Manege, he stages a concert of all the choirs in Liepaja and improvises choirs modified choir samples on his Maschine, creating a unique meta-choir experience. He is also a frequent collaborator with the Liepaja Symphony Orchestra. Anthony made a strange discovery when he checked the local library: he now lives in the area of his ancestor's colonizers. Liepaja is part of Courland, a powerful Duchy in the 17th century. His home, Tobago, was occupied by the Courland army and renamed New Courland. The experience to move unknowingly, via London, from New Courland to the 'old' Courland was transformed into a concept album which he released in 2029. In 2030 the Liepaja/Karosta music hub has attracted worldwide attention but not only for the high-quality music output that is shared as 'creative commons' among the inhabitants. What is unique is the freedom of the musicians to make music without the control of the 'industry'. Independent music is thus redefined on the Latvian West Coast. The dynamism of the music scene, the concerts, the live broadcasts from unplugged performances, live concerts, and recording sessions have also found a worldwide audience in the digital sphere. As opposed to paying for overpriced tickets for Glastonbury, Coachella, Antibes Jazz Festival, or the Ravenna Festival, audiences can come every week-end to 'old Courland' to experience exciting contemporary music. Some artists from the Karosta Collective were seeking out a career in the music industry and left the place. Many came back, either because their short moment of stardom was over or because they missed the community and its creative spirit. After three years in Latvia, Anthony and Jasper are a vital part of the community. The BBC Sounds of 2030 list was just released, without any mention of Toco. But the list has long lost its relevance as an indicator for the latest developments in music. Those who want to get the freshest experience buy tickets to Liepaja.

Index

For Product Safety Concerns and Information please contact our EU representative GPSR@taylorandfrancis.com Taylor & Francis Verlag GmbH, Kaufingerstraße 24, 80331 München, Germany